D Stirling

Simple Propagation

NOËL J. PROCKTER

Simple Propagation

*Propagation by Seed, Division,
Layering, Cuttings, Budding
and Grafting*

FABER AND FABER
London   Boston

First published by
W. H. and L. Collingridge Limited in 1950
New and revised edition first published in 1976
by Faber and Faber Limited
3 Queen Square London WC1
First published in Faber Paperbacks 1981
Printed in Great Britain by
Redwood Burn Limited, Trowbridge and Esher
All rights reserved

© *1958, 1963, 1976, Noël J. Prockter*

British Library Cataloguing in Publication Data

Prockter, Noël J.
  Simple propagation. – New and revised ed.
  1. Plant propagation
  I. Title
  635'.04'3   SB119
  ISBN 0-571-11707-4 Pbk

CONTENTS

7

ILLUSTRATIONS

9

❧ Preface ❧

A question asked over and over again is 'How can I increase my stock of this plant?' In this book I have attempted to answer that question. Nevertheless, propagation is a vast subject and this book is not a treatise for the professional. Rather my intention has been to help the amateur who wishes to perpetuate some particular treasure he has in his garden. The methods I have described, while thoroughly effective, are all of a character which anyone who has patience and ordinary handiness may readily carry out.

I think one reason why more has not been written on this subject is that those who know are inclined to feel that, if the layman is told too much, the nurseryman and professional will be put out of work. But what a delusion they are under. When a woman designs and makes all her own clothes all the dressmakers do not go out of business. Why, therefore, if a man tries his hand at propagating his own plants, should the nurseryman close down?

During two major wars nurserymen should have learned to feel grateful to the amateur for the worthy part he has played in keeping alive valuable plants that might otherwise have been lost. In many cases this would have been impossible had those same amateurs not possessed some knowledge of propagation.

I hope all who read this book will use its information for the betterment and perpetuation of that great heritage, the plants with which our gardens have been so richly endowed by explorers and breeders.

I would like to record my appreciation of all the help which has been given to me in the preparation of this book. In particular I wish to thank the authorities of the Royal Botanic Gardens, Kew, and the Royal Horticultural Society, Wisley, for providing facilities for obtaining many photographs; also Mr. R. Fairman for many helpful suggestions and material from which sketches illustrating this book have been prepared, and Mr. A. W. Mansfield for kindly reading the alphabetical list.

Unfortunately my friend Mr. H. Rofe has not been spared to see the publication of the book in which he took so much interest.

Crawley, 1949   NOËL J. PROCKTER

PREFACE TO FIRST REVISED EDITION

There is little for me to add to my first Preface. In this revised and enlarged edition of *Simple Propagation* I have added the latest information available on this fascinating branch of horticulture. Such additions are air layering, an advancement on the method known as Chinese layering, and mist propagation, an advancement on the sun-frame method. Both of these new techniques are illustrated.

I have also added new genera to the alphabetical list, Chapter Six. With some plants I have discovered new and better methods of propagation and these have been included. In some instances where plant names have been changed, these have been brought up to date and cross-referenced.

Crawley, 1958   NOËL J. PROCKTER

PREFACE TO SECOND REVISED EDITION

In 1958 I revised *Simple Propagation*, which was first published in 1950, and a second revision has now been made necessary.

I have, in Chapter Five, brought the fruit stocks up to date by the inclusion of the Malling-Merton rootstocks, abbreviated to MM.

To the alphabetical list, Chapter Six, new genera have been added, with further cross-references.

In the first revision I added the modern techniques of air layering – an advance on the ancient method known as Chinese layering – and of mist propagation, and improvement on the sun-frame method; both are fully described and illustrated. In this latest revision certain other illustrations have been replaced.

Crawley, 1962   NoËL J. PROCKTER

PREFACE TO THIRD REVISED EDITION

It would appear from the many requests I have had for a new and revised edition of *Simple Propagation* that it is still much needed by amateur gardeners.

In this third revised edition, the East Malling fruit stocks have been brought up to date, and in Chapter Five, with the invaluable help of Dr. B. H. Howard of East Malling Research Station, I have added Chip Budding.

Many advances in plant propagation have occurred since the first edition was published in 1950. Apart from such techniques as the use of hormone plant growth substances, mist propagation and air layering already included in previous editions, we have had such advances as soilless composts, polystyrene propagators, and electric and gas steam soil sterilizers, all of which considerably help the amateur and professional propagator alike.

Finally, in Chapter Six the number of genera has been increased to over 550 different plants; among the additions are included house plants which have become so popular in recent years.

Crawley, 1975   NoËL J. PROCKTER

❧ Acknowledgements ❧

In this new and revised edition of *Simple Propagation* I am particularly grateful to Denys Baker for his pleasing and clear descriptive line drawings. I am equally indebted to Peter Wood, Editor of *Amateur Gardening*, for the use of many of the photographs in this edition.

To Dr. B. H. Howard I am grateful for his help and advice with the text on chip budding, and to East Malling Research Station for use of photographs.

To Arthur Hellyer I owe special thanks for all the help and advice he has given me in all the editions of *Simple Propagation*.

My thanks also to the following who have allowed me to use photographs, etc:

Camplex, Robert J. Corbin, Brian Furner, John Galbally, Nobles (Wellingborough) Ltd., Shilton Garden Products, Wright Rain Ltd., and to Messrs. Allen & Unwin for permission to use material from *Seed and Potting Composts* by W. J. C. Lawrence and J. Newell.

Last but by no means least, my appreciation and thanks for the help and co-operation given to me by my publishers.

❀ 1. Seed ❀

Seed is the means by which the majority of flowering plants spread and reproduce themselves, but by no means the only method. As examples of other means, one might mention the blackberry, which layers itself, thus producing a new plant vegetatively; also the narcissus, new plants of which are produced from young offsets. And in the case of some lilies new plants are even formed in the axils of the leaves (i.e. where the leaf stalk attaches itself to the main stem); these are termed bulbils.

Every seedling is an entirely new plant, and though each seedling will have the main characteristics of the parent plant, in many cases it may differ considerably in important details. It is largely through this variation in seedlings that we have got the tremendous variety of colour and form in our garden plants. All the multitudinous named varieties that beautify our gardens today, have not been produced purely by chance. Many great gardeners, professional and amateur, have spent their lives in raising new and better varieties from seed and so improving the colour and form of some particular plant or the flavour or quality of a fruit or vegetable.

Look at the improvement that has taken place in such plants as the delphinium, lupin, dahlia, rose, apple, sweet pea, and the great variety of vegetable and annuals we can now choose from in the nurseryman's and seedsman's catalogue. The potential variability of the seedling, while it can often be of great service to the gardener, can sometimes be a nuisance. When he propagates a selected variety of apple or delphinium, for example, he naturally requires all the offspring to be identical with the parent. That is one reason why he often has to abandon seed as a method of increase and use some vegetative process such as cuttings, grafts, divisions or layers. Plants so produced are in one sense not really new individuals at all but simply extensions of the original plant, resembling it in every particular.

Sometimes it is possible to overcome the natural variability of seedlings to an extent sufficient to satisfy garden requirements. This has been done with almost all vegetables and most popular annuals such as sweet peas, French and African marigolds, zinnias and stocks. To get these perfections seedsmen go to great trouble to rogue and re-rogue acres of plants when in flower, to see that the degree of variation, which is bound to occur even in such 'true breeding' selections, is not beyond what may be permitted without annoyance to the gardener.

Hybrid vigour. In the past ten years seedsmen have been offering seeds which are special hybrids and have attached to them the symbols F_1 and F_2 hybrids. To produce these hybrid vigour varieties, two selected parents have to be crossed, which in turn create an F_1 hybrid, i.e., a first year hybrid. With these hybrids new crosses have to be made each year in order to produce fresh seed, as without going through this procedure the resultant seed would not be true to type. In fact it is better not to use seed saved from F_1 hybrids as there is no guarantee that it will come true. The results of these crossings have vastly improved colours, shapes and sizes and given longer flowering periods and better vegetables. An F_2 hybrid or a second year hybrid is the result of F_1 hybrid plants pollinating themselves naturally, from which the F_2 hybrid seeds are saved for the following year. In either case the work entailed is both expensive and time consuming and it is for this reason that these hybrid vigour varieties are more costly to buy.

Plants that have been improved through hybrid vigour include hardy and half-hardy annuals and biennials: antirrhinums, begonias, calceolarias, marigolds, pansies, petunias and zinnias; vegetables: cucumbers, melons, sweetcorn and tomatoes.

Harvesting and storing of seeds. Before I come to the requirements and methods of sowing seeds, it is well that the reader should know something about their harvesting and storing. For I am certain that many amateurs will like to try their skill at saving their own seeds and even making a cross or two to see if they can produce something quite new. If the seed that is to be sown has not been harvested or stored correctly, germination will be the poorer for it.

When choosing plants for seed, pick out those that are clean, healthy and as true to character as possible. Choose a dry or sunny day and gather the seed-heads or pods just before they begin to open and disperse their contents. If the whole plant is nearly matured it may be pulled up complete and hung up in a paper, not polythene, bag. Severed seed-heads and pods may be spread out on sheets of paper and placed by a window or on the staging in a greenhouse, so that sun and air may complete the ripening.

Once the seed-heads are dry, the seeds may be shaken or rubbed out and cleaned by being sifted through sieves of different sizes; or place them in a scoop from a pair of scales in which they can be rocked backwards and forwards, and gently blow on them to remove dust and other fragments left behind from the old seed-heads – this is best done outside on a calm day. After the seeds are thoroughly clean they may be packeted up in paper bags or envelopes and stored in a dry cool room where the temperature is as constant as possible. All seeds should be labelled clearly and correctly: I cannot emphasize this too strongly, as there is nothing more irritating than to look through seeds or plants and say 'I think this is so and so, but I am not sure'.

Sealed Seed Packets. The first people anywhere in the world to introduce a hermetically sealed seed packet were

Bodgers in the United States of America, about 1960. This type of packaging ensures a long packet life for the seeds – as much as five years in some cases. The seed must be of good quality and it must be dry – as little as 4% to 6% moisture level – and the air in the packet must also be dry, which necessitates packaging in a controlled humidity chamber or room. The packets themselves are usually of plastic with an aluminium foil lining the essential being that they must be vapour proof.

The first to use this process in England was Dunns under licence from Bodgers in late 1963 or early 1964. Then Hurst introduced Life-Pak in 1964 for the seed trade and market gardener. Then in 1966 Suttons launched Harvest Fresh for the retail trade and amateur gardener. And in 1975 Hurst introduced a So-Fresh retail pack.

Once the pack is opened the seeds very quickly come back into equilibrium with ambient humidity which means probably that their moistures would rise to 10% to 14% depending on species. This is why when the packet is opened the ageing process of the seed starts again immediately.

Pelleted seeds. The introduction of pelleted seeds in 1970 has some advantages for the amateur, both in saving seed and in reducing the irksome task of thinning seedlings.

Small seeds when in pelleted form are easier to handle and plants can develop without checks when thinned or transplanted. The pellets when in contact with moisture disintegrate, which starts germination.

TYPES OF SEEDS

It is not my place in a book such as this to weary the practical gardener with botanical terms or knowledge. But it is right that a few words should be written about the different types of seeds which the amateur is likely to meet. Broadly we may define six classes, judged from the standpoint of their special cultural requirements. They are as follows:

    (a) Dust-like – begonia, meconopsis

(b) Hard-coated – rose, sweet pea, walnut, peach, oak
(c) Fleshy – broad and runner beans
(d) Oily – magnolia, castor oil plant
(e) Winged – sycamore, ash, lime
(f) Plumed – scabious, erigeron

(a) **Dust-like seeds.** These often lose their germinating powers rather rapidly and should in general be sown as soon as is practicable after they are fully ripe. Even more important is the fact that they can easily be killed by overdeep sowing. Very often it is best not to cover them at all but to sow on the surface of very finely broken soil, pressing them gently into this with a piece of glass or a very smooth and dry wooden block and then covering the whole seed pan or box with a pane of glass and a sheet of brown paper, to be removed as soon as germination occurs.

(b) **Hard-coated seeds.** Seeds that are hard-coated often want rather special treatment. Take for example the sweet pea. Certain varieties, particularly the crimson and maroon coloured ones, have such hard coats that they germinate badly unless chipped with a sharp knife. Be very careful to chip on the opposite side of the seed to that of the 'eye'.

Many amateurs like to raise their own rose stocks, hawthorns for hedges, and peach trees from stones. And I might say that I have often heard of amateurs having good crops of peaches off trees grown from peach stones, though to enjoy such prizes means patience – at least seven years will elapse before they bear fruit.

All these seeds that I have mentioned, and many others, require stratifying, a process by which the hard coat is softened. To achieve this, holly berries, peach stones, rose hips, etc., are placed in pots, pans or boxes filled with a mixture of half sand and half peat (1) and plunged in a bed of ashes or soil on the north side of a hedge, fence or wall. Pots and boxes need to be covered with a slate or tile to prevent damage by mice – small-

1. To stratify seeds place them in a pot or pan filled with a mix of half peat and half sand.

2. Pots or pans plunged in soil or ashes and protected by wire netting from birds and vermin.

meshed wire netting can also be used as protection against vermin and birds (2). Here they remain, six, twelve or eighteen months before they are ready for sowing, and the more they are frozen during the process the better. When the time comes the seeds may be sifted through a fine sieve, or washed and sifted, or sown direct. With small seeds, it is wiser to wash them, as then the seed when dried (slightly), can be mixed with dry sand and sown more evenly.

(c) **Fleshy seeds.** Seeds, like beans and garden peas and even marrows, germinate quicker if soaked in water for forty-eight hours before they are sown. And marrow seed is best kept two years in store, as better crops will be the result.

(d) **Oily seeds.** These lose their vitality very rapidly. It is a wise plan to try and sow as soon after gathering as possible.

(e) **Winged seeds.** In this group the wings may be removed when cleaning and then the seed sown in the normal manner.

(f) **Plumed seeds.** I have specially mentioned scabious as an example of a plumed seed because of the failures some raisers have if the seeds are buried too deeply. With choice varieties it is really worthwhile to place each individual seed separately in the soil, leaving its plume sticking above the surface.

SOWING

Requirements for germination. The essential requirements for the successful germination of any seeds are moisture, air, warmth and, eventually, light. It is only when the first three conditions are provided that germination begins. As long as the seed is kept dry and at a low temperature germination will not commence.

Moisture is required in order that the outer coat of the seed may be softened, so that the root and seed leaves can push their way down into the soil and up into the air, and also, of course, to supply the rapidly multiplying plant cells. Warmth quickens the life process, and the precise degree required will depend upon the character of the plant, plants from the tropics requiring more than those from a temperate region. Too much heat may result in very rapid germination and a weakened seedling, which will either dry up or be overcome by fungus, the dreaded damping-off disease, which is described on page 29. Air, and particularly the oxygen it contains, is absolutely essential for the biological and chemical changes that are going on inside the seed and seedling. Nevertheless the importance of air is often overlooked and failure to ensure adequate aeration accounts for many disappointments. Lastly, there is the question of light. Without this the seedling cannot become strong and grow into a healthy plant to flower and produce more fruit and seed.

3. Large seeds like lilium can be spaced out evenly and covered with their own depth of soil.

Seeds vary considerably in size, and it is partly by this variation that we are able to classify them as to how they should be sown and under what conditions. A point often raised by the amateur is at what depth seeds should be sown. The old rule can still be followed, namely that of covering them with their own depth of soil (3). This means that very small seeds are scarcely covered (the smallest not at all), whereas seeds like holly need to be 1 cm ($\frac{1}{2}$ in) under the soil, and horse chestnut as much as 2·5 cm (1 in) deep. But there are exceptions (as in most rules), for garden peas should be sown 5 cm (2 in) below the soil, and broad beans 7·5 cm (3 in).

For the amateur's guidance and convenience it will be sufficient, I think, if I consider the sowing of seeds under the following three headings:

I   Those seeds that can be sown direct into the open garden or bed.
II Those seeds that require a cold frame or cold greenhouse.
III Those seeds that require a warm greenhouse or warm frame, such as an electrically heated propagator.

Some seeds are best sown as soon as ripe, but for the majority it can be said that those in Group III are best sown from the end of January to early March, with certain exceptions, e.g. *Cyclamen persicum* is often sown in July or August. Seeds in Group II are mainly sown in March and April. For Group I the season is a long one – April to August, and in favourable seasons even as late as September, but August as a general rule is late enough. As I have mentioned, some seeds must be sown as soon as ripe, and this usually means late July or August.

I OUTDOOR SOWING

In sowing seed in the open garden or border the sower must be guided by climatic conditions: it is better to wait until the soil is of the right temperature and in good workable condition, rather than to keep strictly to the calendar. As an example, in

Crawley, Sussex, 8 May used to be known as 'Fair Day', and on this day for generations the local people would sow their runner beans. However, in my view, if the ground and temperature were not suitable on this particular date, it would be better to delay the operation. Experience will soon guide the amateur, for if sowing is carried out under unsuitable conditions, clogging or caking of the soil follows and it becomes so hardened and baked that the seed cannot germinate. The seedsman is sometimes blamed for poor germination which is really due to the weather, or to the wrong time of sowing, or to insufficient preparation of the soil beforehand.

Preparation of seed beds. Whatever you do, make certain that the ground is well cultivated and that you have a good tilth. What is tilth? It is that condition of the soil obtained when it has been worked and reworked through the process of digging, weathering, raking and rolling (sometimes), and by these actions has reached a fine and crumbling state which is ideal for the reception of small seeds. The finer the seed the more the ground must be tilled and broken down.

When making a seed bed I always bring the Canterbury hoe into play, a most useful tool for breaking down the first large clods and lumps. I even give the soil a first rake over with the Canterbury, before finally getting my fine tilth by using a good garden rake, or even a wooden hay-rake.

Methods of sowing. Sowing outside can be done either in rows or broadcast, the latter term meaning that the seeds are evenly scattered over a given plot of ground or in a frame, as opposed to sowing in drills or individually. When sowing in a frame or border, broadcasting is sometimes easier than in the open garden. In the case of the frame or border, the seed may be covered by sifting fine soil over it. For this purpose a flour sieve is suitable. A sieve I have often used is one made by fixing a piece of perforated zinc, as used for kitchen safes, on to the sides of a small box in place of the bottom, and for seeds of moderate size this will be found more useful than a flour sieve.

Before sowing commences, space out the distance between drills with a measuring rod, and mark by sticks. When drawing drills in the open, a garden line and a hoe are required – for small seeds one that is well worn and has a blade coming nearly to a point is best. The hoe should be kept close up against the line. For larger seeds, e.g. garden peas, broad beans, etc., a larger hoe or Canterbury hoe is required and a flat broad drill is drawn out. Some gardeners prefer to use a spade.

It is always a problem to know how much or how little water should be given to seed beds. As a general rule watering should not be required when sowing in the open border or vegetable garden; but if the ground is very dry, water the drill after opening it up and before sowing, making certain that the water has drained away before the seed is sown. In hot dry weather when the ground is really warm, this has a marked effect on germination.

Once the seeds are in position, the next operation is to cover them in by drawing back the displaced soil with a rake. This is a job that takes care and understanding, an action of gentleness and ease, enough movement to cover the seed to the required depth but no more. It is best done by drawing the rake along the top of the drill end to end, holding the rake in such a position that it just touches the soil. If raked from side to side, the seeds are likely to be disturbed. This, however, does not apply in the case of the larger seeds like peas, as there is more soil either side of the drill to be replaced.

Again I say, once the seeds are sown, do label them clearly, also adding the date; this is a help in future years, for one can correct errors in timing.

Pelleted seeds. For pelleted seeds, the drills or seed bed must be watered before sowing, especially if the soil or weather is dry, as moisture assists disintegration and hastens

SEED

germination. Also during dry weather a sprinkling of water can be given after sowing, and at no time during the germination period should the soil be allowed to dry out.

II SOWING IN FRAMES

Seeds that require a cold frame are sometimes sown in pots, pans or boxes, and sometimes in the soil bed itself, particularly in the case of early vegetables like carrots, radishes, lettuces, spring onions, etc. By using pots, pans or boxes, greater individual attention can be given, especially with regard to watering and transferring to another frame when germination takes place.

Some seeds, particularly those of certain alpines, trees and shrubs, take as long as two or even three years to germinate. A good plan in winter is to place the pots, pans or boxes on the north side of a fence or hedge and leave them out in the open quite uncovered, allowing the snow and frost to play nature's part. This will do no harm, and by the following spring germination will be markedly more rapid. It is, however, as well to cover the containers with some form of netting to keep off birds, mice and other rodents.

III SOWING IN GREENHOUSES

Seeds that require a warm greenhouse in which to be raised are usually sown in pots, pans or boxes, for the same reasons of convenience as those in Group II. Amateurs often get the idea into their heads that the greater the heat the better will be the result, but the outcome in many cases will be the reverse. I will not be too dogmatic about temperatures, but over many years of experience propagators have found that most half-hardy annuals germinate best in a temperature of 16°C (60°F), while seeds such as those of cyclamen, gloxinia and begonia germinate better in a temperature of 18–21°C (65–70°F).

A point I have not touched on very fully is the effect of

26

light and shade, after the seed has been sown. Many seeds germinate more rapidly in the dark, though light is essential for the seedlings once they appear through the soil. For this reason it is common practice to cover seed pots, pans and boxes with a sheet of newspaper, or if a greater density is required, brown paper may be used.

Also, it is usually an advantage to have a sheet of glass covering the seed pan, before the sheet of paper is placed over to exclude the light. The purpose of the glass is to slow down evaporation and reduce the need for frequent watering.

Preparation of pots, pans and boxes. Pans or boxes suitable for the germination of seeds can be purchased, and ready-made boxes are usually 36 × 21 × 7·5 cm ($14\frac{1}{2}$ × $8\frac{1}{2}$ × 3 in) inside measurements. But these can be made at home quite easily, or suitable old boxes may be used. If constructing the boxes yourself try and purchase seasoned wood, and use galvanized nails. Where seed boxes are made of wood, a dressing of a wood preservative, such as horticultural 'Cuprinol', will prolong their life very considerably.

Whether pans, pots or boxes are used, even when made of plastic, good drainage is essential. Also, cleanliness should be an unwritten law. See that your pots, pans, boxes, crocks, frames and greenhouses are clean and free from disease germs, as far as is humanly possible. I do not wish to be a 'Job's Comforter' by stressing too much the prevalence of disease, but it makes the work of a propagator much easier if all reasonable precautions are taken to ensure hygienic conditions. There is no advantage in using sterilized compost if the containers are not clean. See that you have good drainage: sufficient crocks (a large one over the main hole) or clinkers, and a covering of coarse siftings from the compost. Then on top fill in the seed compost itself, using the tips of the fingers and firming the corners and sides first, otherwise these can be overlooked and the compost surface then tends to slope down to the sides, leading to drainage troubles and so on. Finally

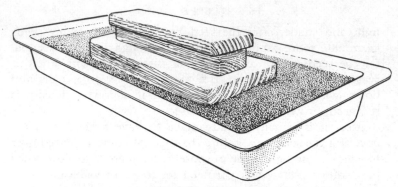

4. A wooden patter used to level off soil in a polystyrene seed box.

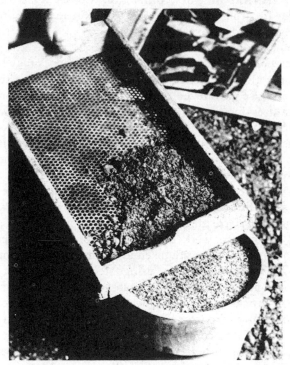

5. Covering seed after sowing with fine compost, using a home-made perforated zinc sieve.

finish off with a patter or the bottom of a pot (4). When using a soilless compost, crocks are not needed for drainage and only a light tap of the pot or pan on the bench is necessary, or finger firming when using boxes.

After the pans, pots or boxes have been filled with compost, water will be required for successful germination. This is best supplied by placing the container in an old bath or sink with the water level coming to just below the rim of the pan or pot, and to 1·2–1·9 cm ($\frac{1}{2}$–$\frac{3}{4}$ in) below the edge of a seed box, and then gradually letting the water percolate through. Once the water has reached the surface, the container should be taken out and allowed to drain, and when thoroughly drained some fine silver sand may be sprinkled over the top and the seed sown and covered with sand or fine compost, using a sieve (5), to a depth determined by its size and nature, as I have already described. Containers filled with a soilless compost can be watered on top to settle the compost. Water finally with a *very fine-rosed* can and cover with a sheet of glass and paper, or paper alone. As soon as the seed has germinated, the glass and paper should be removed, and the container placed where there is ample light to prevent the young seedlings becoming yellow and spindly, or valuable stock may soon be spoilt.

When seedlings are large enough to handle, prick off into pots, pans or boxes.

Damping-off. This bogy of the propagator, particularly when he is raising plants from seeds, is not one single disease, but is caused by a group of closely allied fungi. These attack seedlings and sometimes older plants at or just above soil level, causing the conditions known as Damping-off, Collar Rot or Foot Rot. The one we are most concerned about here is Damping-off. It is easy to detect, as the stem will be seen to shrivel, starting at the base, and discolour, and the plant wilts and soon topples over. Large batches of seedlings can collapse in a few hours. This fungus flourishes under damp, stuffy conditions (hence its name) and seedlings that are most prone to attack are those that have become drawn and spindly through lack of light and air.

Fresh soil should be used, or old soil sterilized with steam or with formalin. Seeds should be sown thinly, and the seedlings pricked off as early as possible. Ventilation must be as free as possible, and overwatering avoided. If, despite precautions, the disease appears, remove and burn all infected seedlings, taking with them a little of the surrounding soil. The remaining seedlings can be watered with Cheshunt Compound fungicide. This can be made up at home or bought ready mixed. The formula is:

Copper sulphate (finely ground) 2 parts by weight

Ammonium carbonate (fresh) 11 parts by weight

Mix the two chemicals thoroughly and store in a stoppered glass jar for at least twenty-four hours. Dissolve 28 g (1 oz) of the mixture in a little hot water and make up to 9 l (2 gal) with cold water. Use the solution at once. It may be watered on with a fine-rosed can. It is wise to water with this fungicide before sowing as a preventive; or a captan fungicide or proprietary seed dressing may be used.

SEED AND POTTING COMPOSTS

It is as well to realize that once the seeds are sown in the soil they have their life before them, and to a great extent their well-being rests with the propagator.

If the soil or compost in the containers in which these seeds are sown is not of the right nature, or if the propagator has firmed the soil down to such an extent as to form a cake through which the seedling has to struggle, we are not giving it a fair chance. But today, thanks to scientific research, there is a seed compost available, known as the John Innes compost, which removes many of these pitfalls, and there are also the soilless composts.

At the John Innes Horticultural Institute, an exhaustive series of experiments was made over forty years ago which proved that if certain requirements are met when preparing composts, good results can be guaranteed. These are the three main requirements:

The compost must contain sufficient food for the needs of the plant.

It must be able to hold sufficient moisture and air without becoming waterlogged.

Finally, it must be without the many organisms which cause such afflictions as damping-off.

Provided these requirements are met, most of the uncertainties of seed raising and producing young plants are swept away.

Compost materials. A good compost consists of loam, moss peat and sand. Of these three, nearly all the plant food is supplied by the loam, whereas the peat absorbs moisture and keeps the compost open, and the sand ensures free drainage. To obtain the best results from turf loam, it must be stacked for some months before use. And I cannot overstress the point that the peat must be of horticultural grade, neither too fine nor too rough, and granulated. The sand, too, must be coarse.

None of these three ingredients, however, contains sufficient plant food to produce good seedlings. But if these materials are mixed in correct proportions, they form the basis of both seed and potting composts. As loam is the only ingredient containing 'food' of immediate value to the plant, fertilizers must be introduced to make good any deficiencies, and I therefore give the formulae of two composts which are still popular – they are known as the John Innes composts:

COMPOST FOR SEED SOWING

2 parts loam (sifted through 3·7 mm ($\frac{3}{8}$-in) sieve)

1 part moss peat (horticultural grade)

1 part coarse sand (grading evenly from dust to 1·3 mm [$\frac{1}{8}$-in] particles)

Add to each 36 l (1 bushel) of the mixture:

42 g (1$\frac{1}{2}$ oz) superphosphate and 21 g ($\frac{3}{4}$ oz) chalk.

COMPOST FOR GROWING-ON
  7 parts loam (sifted through 3·7 mm ($\frac{3}{8}$-in) sieve)
  3 parts moss peat (horticultural grade)
  2 parts coarse sand (grading evenly from dust to 1·3 mm ($\frac{1}{8}$-in) particles)
Add to each 36 l (1 bushel) of the mixture:
  21 g ($\frac{3}{4}$ oz) chalk or ground limestone
  113 g (4 oz) compound fertilizer, which is itself prepared by mixing equal parts by weight of hoof and horn meal and superphosphate of lime, and half a part of sulphate of potash.

0·9 kg (2 lb) hoof and horn, 0·9 kg (2 lb) superphosphate of lime and 0·45 kg (1 lb) sulphate of potash, will make a convenient quantity of the base fertilizer, which can be stored in a tin indefinitely.

These two mixtures are suitable for all types of seedlings and plants without any modification whatsoever. So far there is no proved reason for using different mixtures for different plants.

Alternative materials. Although the fertilizers given have proved the best for their respective purposes, others may be used.

Dried blood may be used instead of hoof and horn meal and at the same rate, and muriate of potash at 14 g ($\frac{1}{2}$ oz) per bushel in place of sulphate of potash. As a substitute for peat, well rotted oak or beech leaves may be used. But as good loam is the basis of compost, every effort must be made to obtain it.

Soilless composts are rapidly taking the place of those made from loam, peat or leafmould and sand, and the John Innes type composts which have been in general use for the past forty years. As loam is becoming more and more difficult to obtain, the soilless composts are now in general use, both commercially and by amateurs. Nevertheless, John Innes type composts can still be bought but see that they are obtained from a reliable source.

There are today several brands of soilless composts on the market, which I give in alphabetical order: Arthur Bowers, Baby Bio, Croxden, Kerimure, Levington and Vermipeat; no doubt there are and will be others, but all are based on peat. The advantages of peat over loam are several: it does not usually contain any weed seeds, its decomposition is slower than soil, it is practically sterile and therefore is more likely to be free of disease organisms, and it has the great advantage of holding water as well as air, which means it does not dry out so quickly, yet does not easily become waterlogged. Although its nutrient content is comparatively low, it is an easy matter to adjust this by adding fertilizers.

In the New Levington Compost introduced in 1975 a special wetting agent is added giving it improved water absorption qualities and ensuring less wastage of water as it does not drain through so rapidly and therefore cuts down on watering.

The last of these soilless composts I have mentioned, i.e. Vermipeat, contains vermiculite, which contains up to 8 per cent of potassium and magnesium, together with essential trace elements.

Many soilless composts contain a percentage of sand. All should be properly moistened, either before or after filling the pans, pots, boxes, etc., but preferably beforehand. The compost 'Plantgrow' has sand added which makes it a useful general soilless compost.

Composts of increased strength. It has been found that certain plants, particularly the more vigorous, can take increased quantities of fertilizer and chalk with advantage, as they are potted on into larger sized pots.

Thus for begonias, calceolarias, carnations, chrysanthemums, cinerarias, cucumbers, cyclamen, hydrangeas, marrows, pelargoniums, *Primula malacoides*, *P. obconica*, tomatoes, schizanthus and strawberries the quantities of base fertilizer and chalk can be doubled when the 11-cm (4½-in) pot size is

33

reached, and tripled when a 20-cm (8-in) pot or larger is needed.

Soil sterilizers and sterilizing. When sowing seed, clean soil free of fungus disease, pests and indeed weed seeds, certainly helps in successful germination, provided the seed is in good condition. If the peat and sand used are clean, these will not require sterilization, but the loam will, and also leaf soil if used in lieu of peat. Materials must be sterilized separately.

Sterilization by steam is the most satisfactory method, as it does not dry the soil, or burn it as in the case of sterilizing by fire. For small quantities of soil, the following is a convenient and efficient method of sterilizing, using a 3·3-l (6-pt) saucepan.

'Saucepan method (6 pints).[1]

'Sieve the soil through a $\frac{1}{4}$ or $\frac{3}{8}$ in sieve, spread the soil out thinly, under cover, for several days until it is approaching dust-dry. Take the saucepan and pour into it a measured $\frac{1}{2}$ pint of water and heat rapidly on gas or electric cooker; immediately water boils, fill saucepan with soil to within $\frac{1}{2}$ in of the top; put on lid and using a "minute timer" continue boiling for precisely 7 minutes. Remove saucepan (without taking the lid off) and stand for a further 7 minutes. Pour out soil thinly on to a clean surface to cool. When it has cooled mix it into a compost, or alternatively, store in a clean container until required. Do not allow the soil to dry out again and, if necessary, moisten it a little with tap water.

'The saucepan method is safe and efficient because the amount and depth of soil are small and steaming can be controlled with ease. A 3-pint saucepan can also be used, in which case the boiling time is reduced to $3\frac{1}{2}$ minutes but the standing time of 7 minutes retained as before.'

Although the saucepan method of soil sterilizing is successful there are several efficient sterilizers on the market today which

[1] W. J. C. Lawrence and J. Newell, *Seed and Potting Composts*, Allen and Unwin, 5th edn., 1962.

6. Nobles Steam Soil Sterilizer. For each 0·014 cu m ($\frac{1}{2}$ cu ft) of soil sterilized, $\frac{3}{4}$ of a unit of electricity is used.

7. Shilton Gas Heater and Steam Soil Sterilizer, on bench; this will sterilize 0·014 cu m ($\frac{1}{2}$ cu ft) of soil.

the amateur can use. There is the Nobles Soil Sterilizer (6) with a bucket system which sterilizes $\frac{1}{2}$ cu ft of soil by steam and is powered by an automatic electric element. There are other makes which are powered by electric immersion elements. These soil sterilizers are chiefly made in sizes that take $\frac{1}{2}$ cu ft of soil, though there are others which sterilize one or two bushels. All these sterilize by dry heat. Steam sterilizing is to be preferred. Finally, there is the Shilton Gas Sterilizer which is operated off the Shilton Automatic Gas Heater, and is a steam sterilizer (7). Manufacturers' and suppliers' addresses can be found on page 239.

Mixing the compost. If the soil is dry before sterilizing begins, it will be right for mixing and using as soon as it has cooled. The sterilized loam must be spread out in a flat-topped heap; add the peat and sand and on top of these spread the small amounts of fertilizers evenly. To ensure thorough mixing turn three or four times.

If John Innes composts are properly prepared, and used under clean conditions, the little extra work entailed is well repaid, for not only are weeds eliminated, but the danger of damping-off disease is removed. Provided normal attention is paid to watering, the growth will be more rapid and better balanced than in carelessly prepared mixtures. I can guarantee that the extra trouble taken will be handsomely rewarded by excellent results.

PRICKING OFF AND HANDLING SEEDLINGS

It is the aim of a good propagator to prick off his young seedlings as soon as they are large enough to handle. This is usually when the seedlings have grown their seed leaves, that is, the first leaves to appear. These seed leaves are usually much simpler in form than the true leaves which appear a week or so later.

These young seedlings should be treated with utmost

respect, and cared for as a mother does her baby. Again I say, use the John Innes compost or a soilless compost for filling up the pots or boxes for pricking out. For very delicate seedlings use a small piece of label or a flat wooden ice cream spoon with a V notch at the end in which the seedling can be picked up, to save damaging the tender stem and seed leaves **(8)**.

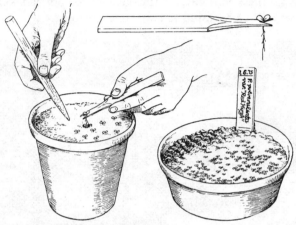

8. Pricking out seedlings with a V-notched stick. Top: the V-stick with seedling in V. Transferring a seedling (left) to a hole made with dibber – *Note* how seed pan (right) is clearly labelled.

The containers into which they are to be pricked off should be clean and dry, for if they are wet and dirty it will be found difficult to remove the young plantlets when they are required for potting on or planting out into bed or border.

Also see that your pots or pans, etc., are well crocked, as in the case of seed containers, and likewise covered with riddlings, coarse ashes or small clinkers, as perfect drainage is still of primary importance. Crocking is not necessary, however, when using a soilless compost. In the same way as before, proceed to fill up with the compost, using your fingers to do this, and finally smooth over with a flat piece of wood and pat down with a patter – or use the bottom of a pot.

The soil should be of the right consistency so that watering will not be necessary. If, however, you do require to water, this should be done with a fine-rosed can, and the water applied some little time before the seedlings are to be pricked off, to enable the boxes or pots to drain thoroughly.

With a soilless compost, a light shake of the pot and watering is all that will be required to settle the compost.

9. Pricking out *Primula malacoides* seedlings from a seed pot into a seed box.

Always keep a box of dry silver sand close at hand, and sprinkle some on the surface of the soil before starting to prick off.

Separate the seedlings out with as little damage to the roots as possible and then replant them singly in their new quarters, making sure that the roots are dropped well down into holes prepared with a small wooden dibber. Firm the soil around the roots with dibber or fingers (9) and, when the pan or box is filled, water thoroughly through a fine-rosed can to settle the soil further around the roots.

Where seedlings of achimenes, begonia, gesneria, gloxinia, saintpaulia and streptocarpus, or similar plants which have a dust-like seed, are concerned, prick them out in little clumps, as to prick them out individually would require infinite patience.

Label correctly and carefully, and also put the date on the label, so that failures and successes can be recorded to help with future sowings and transplantings; such data are invaluable to the really keen propagator.

Once the seedlings have been pricked off, labelled and watered, shade them for the first few days with paper.

Never allow the soil to become too wet or too dry, for such changes are asking for trouble. Once seedlings become hardened they may be checked seriously and in some cases ruined.

As soon as the seedlings are large enough, they must be kept on the move by being potted on or planted out, but here I will leave seed raising, for the next step is cultivation, which I do not intend to delve into.

❧ 2. Division ❧

Division hardly requires a definition, for the term is really self-explanatory. It means precisely what it says, namely that some plants can be divided into several pieces, each of which will grow on into a complete new plant. Good examples are the Michaelmas daisy, golden rod, the gladiolus corm, the narcissus bulb and the Madonna lily (*Lilium candidum*), and many shrubs.

With a great many plants this simple form of vegetative propagation provides so satisfactory a method of increase that no other need be considered. Although division is so simple, it is not always carried out with success, and some skill and knowledge are needed.

The plant required for division must be of a kind that makes numerous shoots or offsets, from either above or below ground level. Consider, for example, a shrub or tree with a single basal stem, such as the familiar Christmas tree or common spruce (*Picea abies*), or a carrot or parsnip, with a solitary crown bud from which all leaves grow. Such cannot be increased by being pulled apart into several pieces.

Ideal subjects for division are rather diffuse plants such as I have already mentioned – Michaelmas daisies and golden rod. These make such a tangle of shoots and roots that it is simple to pull or chop out a portion, and separate it into as many pieces as can be found with roots attached. Each individual piece can then either be planted or potted to form a fine new specimen in a matter of a few months.

It would be impudent of me to imply that division is always as simple as this.

CLASSIFICATION OF MATERIALS

In order that the reader may be able to determine quickly and easily how any particular type of plant may be divided, I have split suitable plants into six groups, taking into account their natural mode of growth. These six groups are as follows:

(a) Those plants that are naturally spreading and free rooting in their habits, e.g. the Michaelmas daisy. This is the easiest group.

(b) Those of a more compact nature, and often provided with plenty of fleshy roots or even woody crowns, which are usually plentifully supplied with growth buds, e.g. the lupin and delphinium.

(c) Those with rhizomatous roots, really modified stems which creep along at or just below soil level, e.g. the German irises and some creeping grasses.

(d) A large and most diverse group embracing those plants which, while compact and often single stemmed themselves, produce offsets, i.e. complete new plants attached to the parent. Most bulbs and corms come into this group, e.g. the daffodil, tulip and gladiolus. The yucca also produces offsets.

(e) Those plants that are tuberous rooted. In some instances these are increased by cutting up or dividing the tuber, e.g. the potato, or by separating it, e.g. the dahlia or peony.

(f) Those plants which throw up suckers, i.e. new shoots direct from the roots, frequently at some distance away from the parent. In some cases such suckers can become very annoying, e.g. the elm tree, poplar or tree of heaven. Other examples are the raspberry and lilac.

It will be seen from this survey that a very large number of plants can be divided. For the amateur this form of propagation is often all that is required. It is by no means every gardener who can afford to buy plants by the hundred or even dozen, but he can buy one, two, or perhaps three. If their habit of growth falls into line with any one of the groups I have set out, then they can be divided in due course. And to my mind the nurseryman is helped, for the amateur is thrilled by his success and so goes again to the nurseryman to see what new plants he has to offer, even if they do cost 25p, 50p, 75p, or even £1.00 per plant.

Before dealing with each group individually, here are a few general remarks. Do not aim at being greedy. By this I mean, only split or divide a plant up so that each portion when divided is worthy of being called a plant or at least a prospective plant. This applies particularly when dividing new plants.

As a rule, division should take place at normal planting times, i.e. autumn or spring, the latter time to be preferred. It is necessary to divide certain subjects more often than others, because some herbaceous perennials, rock plants, bulbs and corms (but not all), deteriorate after several years in the same spot, and in some cases perhaps die out altogether or become old, coarse and worn out. Many herbaceous plants require dividing at least every three or four years.

Most plants after division should be planted up immediately; this does not necessarily apply to bulbs, as many of these require a rest period which is more often than not best spent out of the ground.

One great advantage which division has over propagation by seed is that for most plants no glass is required. Moreover, stocks are kept true to form and colour whatever their nature.

FORMS OF DIVISION

Although the general principles of division are the same for all plants, methods differ in detail according to the type of growth, and I will deal with these variations under the six headings already enumerated.

(a) **Plants of spreading and free rooting habit.** I suppose there is no better known plant than the Michaelmas daisy,

10. A Michaelmas daisy root being torn apart with the hands.

which may be increased so easily by division. Provided the portion that has been pulled off from the parent has a few small roots or root and a shoot, a fine new specimen will be growing and flowering in a matter of months.

I only use the Michaelmas daisy as an example. There are a vast number of other plants that have quite as diffuse a habit, spreading their roots far and wide so that they can become a nuisance. Golden rod, achillea, *Chrysanthemum maximum* (in a previous garden when I took it over, the border was more than half-filled with *Chrysanthemum maximum*), erigerons, and a number of shrubs, come into this group. Then again, many alpine and rock plants are readily increased by division.

Although many plants can be torn apart by pulling a portion off with the hands (10), this is not the case with every plant. The gardener has to find various methods to divide up his stock. And if the plant to be divided is new, rare or he has fears of losing what he has got, the safest and surest method must be adopted.

11. When dividing tough plants, a pair of garden or hand forks can be used to lever the roots apart. Make sure that the forks go well into the root.

A pair of hands or garden forks may be thrust back to back through a tough clump and used to lever it apart (11). Sometimes I have used a sharp pointed stick to lever between the entangled mass of shoots and roots. A knife may occasionally be required to sever a tough crown.

When using a pair of forks back to back, it is essential that they are placed with care and driven far enough in to reach the base of the plant to be divided; otherwise the result may be that the top will part company from the bottom.

Wash soil from roots if necessary, and with plants such as heleniums and erigerons, comb out the roots with a sharp pointed stick and give the plant a thoroughly good shaking. After such treatment some plants will often practically fall to pieces.

It will be noticed that old overgrown plants become woody, and often very hard. Do not retain the central core, but only the young growths around the outside of the plant. Sidalcea roots become very gnarled, and such plants should have as much as possible of this hard woody core cut off, leaving the younger portion of roots and shoots.

When division has taken place, no time should be lost in dealing with the portions retained. If they are very small, it will be wise to plant them in a nursery bed and allow them to grow on into larger plants. With choice plants I like to take extra care, and I have found it well worth while potting the single pieces or little clumps (whichever is the case) into 7·5-cm (3-in) pots, using any good potting compost, John Innes or a soilless compost, for the job. If potted in autumn they can be overwintered in a cold frame. Give plenty of ventilation on favourable days, and only sufficient water to prevent them from drying out. I have grown many hundreds of herbaceous plants this way.

(b) **Plants with fleshy or woody crowns and with growth buds.** The delphinium may serve as our example of this group. Generally speaking, these plants call for a little more attention than those in group (a).

In most cases a sharp knife will be needed, and whatever you do, make sure that when you cut, the knife is placed accurately and that there is a good sound bud attached to the portion of root about to be severed from the parent plant, for unless the two are attached, you might as well have left the plant undivided. With plants such as the delphinium and lupin, I advise washing so that you can see what you are doing. Use warm water – it won't hurt. Never make a martyr of yourself, for one might just as well do the job in comfort. If the divided portions are small, pot or box up until the young plants are a little more able to fend for themselves.

(c) **Rhizomes.** A rhizome is a shoot or stem, found usually beneath the surface of the soil and growing more or less horizontally. A great variety of plants have rhizomes, some short and thickened, e.g. the German or flag iris (12), others long, thin and straggling, e.g. lily of the valley. Then in some

45

12. The German or flag iris is divided by cutting off a portion of the rhizomatous root with one, two or three fans. The foliage should be cut back to 15 cm (6 in) like a blunt arrow head; the young division is then ready for planting.

cases the shoots or stems become tuberous, e.g. Solomon's Seal, and in plants such as the montbretia both rhizomes and corms are developed. Many grasses have rhizomatous roots. That old enemy of many a gardener, couch grass or twitch, is a fine example of how every little portion can root.

When dividing rhizomes it is as well to discard all unrequired portions, particularly in the case of the iris (12). Only sufficient rhizome should be saved to supply the required food for the newly divided plant. Water-lilies make thick woody roots as much as 10–12·5 cm (4–5 in) in circumference and they are tough roots to encounter. A strong knife is necessary – no dainty pocket knife for a water-lily.

The rhizomatous grasses are very much easier to divide, except that some may be found rather tangled. Not all grasses have rhizomatous roots, and those that have not can be divided as plants in group (a).

(d) **Offsets.** An offset is a complete new plant attached to parent. Such subjects as the narcissus, tulip, crocus, gladiolus, yucca and montbretia are all well-known examples of plants which produce offsets.

Sometimes the offset is produced alongside the parent, as in the case of a tulip or daffodil bulb. With gladiolus the new corms are formed on top of the old parent corms. Incidentally, gladioli produce side offsets as well, much smaller in size than the new corms and often very numerous. The tiny offsets or cormlets may be detached and treated as divisions, but it may be a year or so before they attain flowering size.

(e) **Tubers.** Tubers have some points in common with rhizomes. A tuber is a thickened or swollen shoot or stem, and may be found above or beneath the soil, e.g. the tubers of cyclamen and tuberous-rooted begonias are partly above ground level, whereas the potato, artichoke, dahlia and peony are entirely below the soil.

The actual method of dividing tubers varies somewhat from plant to plant. Potatoes and Jerusalem artichokes may be propagated by cutting them up into several pieces, and provided each portion has an eye, then a new plant can be produced from it.

The dahlia has to be divided by severing the tuberous roots with a portion of stem attached and provided there is a good eye or bud, a new plant will be formed.

Tuberous-rooted begonias can be cut into pieces, but they must first be started off in peat moss litter, so that eyes are produced and the propagator may know where to cut.

(f) **Suckers.** The reason I have placed suckers last in my list of various forms of division is because to my mind they are very closely allied to the layers described in my next chapter. With some shrubs it is difficult to discern the difference between a sucker and a layer. This need not concern the gardener, but it is interesting to note how different methods of propagation are linked up with one another.

The separation of suckers from the parent plant is a very simple and productive form of division. Many gardeners will know only too well how the raspberry sends up suckers a yard away from the original parent canes. All that is necessary to procure new stock is to dig up the newly formed canes. They should be cut off with a sharp pruning knife or pair of secateurs, with a portion of the old root attached and trimmed if necessary, and planted in their new position. This also applies to lilacs, *Rhus typhina* (stag's-horn sumach), *Populus tremula*, and similar subjects.

The young suckers when removed should have any old, woody or unwanted roots and smaller or weak shoots cut off, so that what remains is a strong and healthy young plant, bush or cane.

Suckers are usually taken off at normal planting time, i.e. autumn or spring.

❊ 3. Layering ❊

Layering provides an easy means of increasing a number of flowering and fruiting plants. It is the term used to describe the method of rooting branches, stems or runners while they are still attached to the parent plant. In this way it differs from propagation by cuttings (see Chapter Four), for here the branch or stem is first severed from the parent and then rooted. Layering occurs naturally in many plants, so really the propagator only puts into practice one of nature's methods of reproduction.

Both hard- and soft-wooded plants may be readily increased by layers. In the first class may be mentioned lilacs, roses, loganberries, heathers, rhododendrons, fruit stocks, etc., while of soft-wooded plants examples are the carnation and strawberry.

This form of propagation has a number of advantages. It does not require glass or heat, and it can be practised on quite old plants, though the younger wood must be used.

Some shrubs are best increased by layers, in preference to being budded or grafted. A good example of this is the lilac (*Syringa*). This sends up many suckers, which in the case of a budded or grafted plant will have the character of the stock and not of the scion worked upon it. With a layered plant this is not so and even suckers will be of the named variety. Nevertheless there are certain varieties of lilac which are better budded or grafted on to common lilac, because layered plants produce smaller flowers, e.g. 'Glory of Horstenstein' and 'Massena'. Rhododendrons are better layered than grafted, as *R. ponticum*, the common rhododendron, suckers freely. Plants can also be increased by taking cuttings.

As I have said, this is a natural means of reproduction,

49

e.g. the blackberry which buries the tips of its long, searching growths in the soil. How many times have I been tripped by a naturally layered cane while blackberrying, the 'trap' being partially hidden in the grass. Upon closer inspection and after a few ungrateful words, I have noticed that a young plant has been unseated, and at the growing tip there are a number of white roots. Another good example is the strawberry runner, but this roots at every node or joint. Sometimes one may see natural layering even in the case of branches of large trees such as beech and horse-chestnut, which have swayed in the wind and brushed against the ground. In time the friction from the constant rubbing on the ground causes an abrasion from which adventitious roots appear. These, together with the accumulation of soil and decaying leaves, fix the branch securely and in time it is firmly rooted to the ground.

While this is happening, and new roots are being emitted, the new plant, shrub or tree receives sustenance from the parent. In the garden the new layer is not severed until sufficient roots have formed to sustain the new plant without help from its parent.

Many stocks (but not all) for apples, pears and plums may be propagated vegetatively by layers, and this is preferable to raising them from seed because seedlings are usually very variable, whereas layers will remain true to the characteristics of the parent from which they are taken. Thus in a batch of crab-apple stocks raised from seed no two will be quite alike and some may be far more vigorous or have a far more securely anchored root system than others. But if layers are taken from the dwarf paradise apple known as Jaune de Metz or Malling 9, every one will be alike and will have the same dwarfing and early maturing qualities.

A most useful publication is *Fruit Tree Raising, Rootstocks and Propagation*, Bulletin No 135 of the Ministry of Agriculture and Fisheries, published by Her Majesty's Stationery Office. This deals very fully with all the above-mentioned fruit stocks. Unfortunately it is now out of print, but copies can still be seen in some public libraries or in the Royal Horticultural Society's Lindley Library.

In order that this process of rooting may be speeded up, it is often necessary that the flow of sap be checked in some way. Now there are a number of ways in which this can be effected, e.g. by bending, twisting, making a cut or notch (with a knife) in the portion buried in the soil, or with a ligature similar to that used by a surgeon. The result of this operation is that where the check of sap has taken place, a healing skin or callus forms. It is usually from this callus that new roots are emitted, though occasionally the callus has to be removed and a new one allowed to form before rooting takes place, e.g. *Chaenomeles japonica* (syn. *Cydonia japonica*) and its varieties.

It does not matter which way you check this flow of sap, but generally speaking, twisting or cutting with the knife is most successful; when twisting a shoot both hands should be used, and a half twist should be sufficient to check the sap. When using a knife, a cut is made on the underside just below a bud and half-way through the shoot, so that it is half asunder and the tip of the shoot can be bent very nearly upright.

Almost without exception, all shoots that are layered require securing in the soil by some means or other, though a plant like the strawberry which layers itself, roots readily without assistance. But in gardens they are usually secured in the soil with wire layering pins or wooden pegs, to prevent them from rising out of the soil.

Soil or composts. Whatever kind of layering is being undertaken, it is quite as necessary to take pains in the preparation of the soil around the plant which is about to be layered, as it is when one is sowing seeds or inserting cuttings. It would be unfair to expect a shoot bent down into some rank old clay to put forth young fibrous roots. The ground must be well dug and made as fine as possible by the addition of leaf soil or peat and sand, in order to make a good rooting medium in which to insert the layers.

FORMS OF LAYERING

Methods of layering may be classified under six headings as follows:

(a) Growing tips, e.g. loganberry and blackberry and similar hybrid berries.

(b) Serpentine layering, e.g. *Lapageria rosea*, clematis and vine.

(c) Runners, e.g. strawberry.

(d) Bending down shoots, e.g. lilac, rose and carnation.

(e) Air layering, or Chinese layering, e.g. rubber plant.

(f) Air layering with polythene film, e.g. trees and shrubs.

I will describe the differences in treatment required for these six methods under their respective headings.

13. A blackberry shoot being tip-layered by pegging it into the ground.

(a) **Growing tips.** This, apart from the pegging of runners as in the strawberry, is the simplest form of layering. All that is required is to bend down a loganberry or blackberry cane in the summer and bury the tip 7·5 cm (3 in) covering with loose soil, or if wished the tip can be buried in a pot previously sunk in the ground. In either case, secure the stem with a layering pin or wooden peg (13). Never allow the soil to become dry; in fact it should be kept moist throughout. The buried tip may be severed from the parent plant in November, and dug up or knocked out of the pot and planted in its permanent position or nursery bed. The parent cane will bear fruit the following season if tied to its supports, but the young plant will be two years old before it bears.

(b) **Serpentine layering.** A good example of a plant which lends itself well to this kind of layering is that lovely cool house Chinese climber, *Lapagaeria rosea*. All that is required is to fill a large pot or box with sandy peat, and arrange the growths in such a way that they may be pegged down securely into the compost a number of times, as roots will be formed at each leaf joint or node. When rooted, sever each portion from the parent plant and pot into small pots containing a peaty mixture.

While on this subject I must mention the clematis, the Queen of climbers. Ernest Markham, in his book on this plant, states: 'I consider layering to be the most satisfactory method of propagating clematis'. Plants growing in the open are the most suitable, and strong, two-seasons-old shoots are best, which may be selected at pruning time and shortened to a length of 1·5–1·8 m (5–6 ft), or less if not so vigorous. Sink a 12·5-, or 15-cm (5- or 6-in) pot into the soil (filled with a sandy compost), into which the layers are pegged, or they may be layered direct into the open ground. The layer is prepared by making a slanting cut about 5 cm (2 in) long on the underside of the growth and close behind a joint (14), or the growth given a twist just enough to break the bark; before inserting the cut

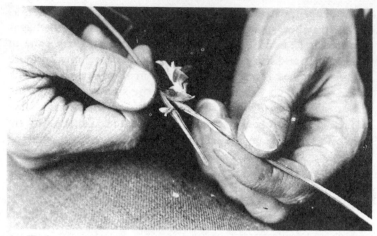

14. To layer a clematis; a slanting cut is made in a stem behind a bud to form a tongue which is later inserted and pegged down into a pot of soil.

or twisted part dust with hormone rooting powder. Each joint may become a plant, but if a really strong plant is required, use only the one at the end of the growth. The layer should be covered with soil and securely fixed with a pin.

By late autumn the newly formed plants may be severed from their parents and planted in the permanent quarters or potted up. By this time there will probably be quite a lot of growth on the young layers; these should be cut back to within 30–60 cm (1–2 ft) of the ground to induce strong healthy plants. Alternatively, layers can be put down during August, being left until the autumn twelve months later, when they may be severed and potted or planted in their permanent position. The clematis can also be propagated by seed, cutting or grafting.

(c) **Runners.** This form of layering is the natural vegetative method of propagation of the strawberry. A word of warning to anyone who proposes to increase his stock of strawberries:

54

he must see that he has a true and healthy stock to propagate from. Today, certified stocks are obtainable from reliable nurserymen. It always pays to have the best, particularly with the variety Royal Sovereign, which is so prone to virus disease.

The ideal time to layer strawberry runners is in June and July. Runners should only be chosen from fruitful plants of a good strain, also free from such pests as red spider mite, aphis and tarsonemid mites. Year-old plants are most satisfactory, and not more than five plants should be taken off one plant; any other runners must be removed in order that the five new plantlets may have the best possible chance. Layer the first plantlet on the runner nearest to the parent, removing all that come after.

Layering can be done in the open ground or in 7·5-cm (3-in) pots. I prefer the latter method. Fill the pots with a loamy soil.

If layering direct into the ground, the soil must be forked over. Place the strawberry runner on top of the soil in a pot (which is sunk in the ground level with the rim) then peg it down with a peg made out of wood, bracken, zinc or galvanized wire (15). Water and syringe regularly during hot dry weather.

15. Strawberry layers can either be layered in a pot of soil or pushed direct into the ground and held in place with a layering pin.

As soon as rooted, the runners connecting the young plantlet and the parent may be cut through. Transplanting is best delayed until a week or so after severing. The young plants are then plunged in some light soil in a cool and shady part of the garden. Here they may remain until wanted for planting out in August or early September. Do not allow them to become dry; also syringe daily to keep down pests.

(d) **Bending down shoots or branches.** Trees, shrubs, roses and bush fruits are layered very extensively by this method – particularly trees and shrubs that are found not to root easily from cuttings. It may interest readers to know that the mulberry can be propagated by layering as well as by cuttings.

When layering a lilac, an azalea, a rhododendron or a magnolia, it is sometimes necessary to form a stool plant, i.e. a plant that has been cut hard back or at least had one or two of its branches cut down so as to produce young and pliable wood with which to work. This is not necessary if there happens to be suitable young growth that can be used.

The usual time for layering is from early March to September as weather conditions permit. Layers of rhododendrons, azaleas and magnolias will need two years before they can be severed from the parent plant.

The ground around the bush or tree to be layered should be well worked, with peat added, and if the soil is heavy, some coarse silver sand also, but do not make the bed too light and spongy, or it will be found difficult to insert the layer firmly in the ground.

For nicking out the V-shaped incision in the soil, a sharp worn-down spade is as useful a tool as any. The depth of this nick should be about 10 cm (4 in). Now take the young shoots chosen for bending down, giving them a half-twist or, if too thick for this, cutting a notch or ring around the stem or splitting with a strong knife **(16a)** or chisel about 60 cm (2 ft) or less (according to the plant) from the growing point. If the

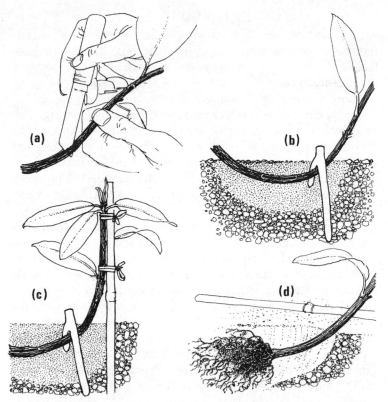

16. Rhododendron layering; (a) first a cut is made in the shoot, (b) after pushing in the wounded portion of shoot into a V-shaped trench it is secured by a wooden peg, (c) the young shoot is finally tied to a small stake or cane, (d) a rooted layer after 18 to 24 months.

stem is split, insert a stone or wooden wedge in it to keep the split open. Then place the notched or split portion of stem in the V-shaped incision in the soil, using a cane to support the end of the shoot in an upright position (16c), filling in with soil and sharp sand and firming with the heel. Secure the layer by using a hazel peg or stout wire peg (16b). These layers may be severed the following spring and transplanted the following autumn, when they will of course be fifteen- to eighteen-month-old plants; however, rhododendrons, azaleas, magnolias, etc., will need two years at least before they are transplanted (16d).

Many amateurs will want to have a shot at keeping an old favourite climber or rambler rose in existence by layering. The preparation of the soil is the same as for the lilac. The shoot or shoots required for layering must have their leaves removed along the portion actually to be inserted in the ground, and about three-quarters of the stem can be denuded. The shoot is now ready to be operated upon. Take a sharp knife and make an incision in a stem on the side nearest to the ground, cutting in an upwards direction nearly half-way through, which will leave a small tongue. Give a slight twist and insert in the soil in a V-shaped slit made as for the lilac but 5–7·5 cm (2–3 in) deep. Peg down with a hazel peg (16b) which I prefer to wire, and cover over with soil, making sure that it is firm. The tongue should then be upright in the soil, and it is from this tongue that roots are formed.

The best time for layering roses is from June to the end of August. Layers may be severed from the parent plant in November if put down early, but generally I would advise waiting till the spring, so as to make sure that they are rooted.

CARNATIONS. The Border and Malmaison types are, as a general rule, layered, whereas the Perpetual carnation, the kind we usually see in the florist's shop, is propagated by cuttings. The latter half of July and the first half of August is the most suitable time for layering Border carnations, as by then the plants will be matured, yet neither too hard nor starved.

Before beginning the job, the following tools and materials should be ready: a trowel, a sharp knife (one with a thin pointed blade is best), and layering pins or pegs made from pieces of galvanized wire 15 cm (6 in) long and bent double in the shape of a hairpin, making each peg 7·5 cm (3 in) in length. Pins or pegs may be purchased from seedsmen and sundriesmen.

Having secured the tools and other equipment, the next essential is compost. A good supply of loam, a little ground limestone well broken up to assist drainage, and a small amount of well rotted cow manure will be needed for this. Or Montagu C. Allwood's mixture recommended in *Carnations and all Dianthus* is excellent: one-third moss fibre litter, one-third maiden loam, one-third fine sharp sand.

Now all is ready. Select the plants you intend to layer, choosing strong, healthy specimens, free from pests and diseases such as aphis or red spider mite. The shoots to be layered are those which have not carried flowers. Choose the sturdiest of these. The soil around the plant will require to be lightly loosened, and the compost placed on top and around to a depth of 2·5–5 cm (1–2 in). If the soil should be very poor, heavy or quite unsuitable, then scoop some out with a trowel before adding the compost, but still make it so that it is 2·5–5 cm (1–2 in) above the soil level. All leaves should be removed from that portion of the shoot which will be under the ground, i.e. generally to within five or six pairs of leaves from the top of the shoot. The leaves may be removed by a sharp upward pull, so that the layer is left clean and without injury. The tongue is made by inserting the point of the knife immediately below one of the joints from which the leaves were pulled off (**17a**). The knife then splits the stem for 1·2 cm ($\frac{1}{2}$ in) or so up through the joint, eventually being brought out at a sharp angle 1·2 cm ($\frac{1}{2}$ in) above it. The pointed end should be trimmed off squarely and neatly so as to form a good tongue. A nice broad base at the tongue will help to encourage better rooting.

Bend the shoot down, making sure the cut is kept well open and that the tongue is not bruised or damaged when pressing it into the compost. This must be carried out gently but firmly. Now place the layering pin or peg in position well below the cut (**17b**). Firmness ensures good rooting.

When all the shoots that are required have been layered, a topdressing of the compost may be given, about 2·5–5 cm (1–2 in) deep. This is followed by spraying with water or

59

(a)

(b)

17. (a) A border carnation shoot prepared for layering; showing the tongue that has been cut.

(b) This portion is pegged into the soil with a layering pin.

watering with a fine-rosed can. The layers must be kept moist throughout the rooting period, which is usually from six to eight weeks. As a precautionary measure dust over with well-weathered soot to prevent maggot. When the layers appear to be well rooted, one may be lifted gently to make certain: experience will soon tell if it is a success. Two or three days before finally lifting the rooted layers, it is beneficial to sever them from the parent plant. After this short interval for recovery from the shock of severance from the parent, pot off into 7·5-cm (3-in) pots or plant out in a well prepared bed.

(e) **Air layering, or Chinese layering.** This is a very old method used by the Chinese, introduced to this island more than 200 years ago. It comes into practice when the branch to be layered is either too stiff or too far away from the ground for other forms of layering to be practised.

The method of Chinese layering is as follows. Either cut off a ring of bark 0·6 cm ($\frac{1}{4}$ in) wide around the stem or make a slanting cut either side of the branch in an upward direction about 2·5 cm (1 in) long, which may be kept open with a small wedge of sphagnum moss. The portion operated on is then bound around with sphagnum moss, or a flower pot is cut in half and clinched together around the stem with wire. Another method of joining the pot is by cementing it. This has an additional advantage in that it makes the pot less porous if a thin layer of cement is smeared all around it. Now fill up the pot with a compost of the type recommended for cuttings. It is imperative that the compost does not dry out, so a bottle or can of water can be suspended above the pot, with a hole in its cork to allow water to seep out and keep the contents of the pot moist.

If the pot is a large one, some form of artificial support will be necessary. When the new plant has rooted, it may be severed from the parent immediately below the base of the pot.

Shrubs may be layered in this way, but it is not often necessary. When it is desirable, a small platform is erected on which is placed a box filled with soil and the branch of the tree or shrub to be layered is inserted in the soil as though it was being layered at ground level.

(f) **Air layering with polythene film.** This is a modern method, and when one considers the advantages, it will be apparent what a wonderful advancement has taken place. This system of propagation is most useful for trees and shrubs. Although the majority are simple enough to propagate from cuttings, ground layers, budding or grafting, there are others that are decidedly difficult. It was Colonel William R. Grove of Laurel, Florida, U.S.A., who discovered in 1947 that the use of polythene film helped the process of air layering so that even difficult subjects could be made to root. Other advantages of air layering are that one can see when the layer has rooted simply by looking through the polythene, and there is no need to water the air layer, as was the case with the old method of using a flowerpot.

THE MATERIALS REQUIRED. You will need a good sharp budding knife, polythene film which should be from 0·10 mm (4/1000 in) to 0·12 mm (6/1000 in) thick (when using tubular polythene, that of 10, 15 or 20 cm (4, 6 or 8 in) wide has been found to be the most satisfactory), sphagnum moss, a hormone plant-growth substance, and self-adhesive tape. Horticultural vermiculite can, if wished, be used in place of sphagnum moss, or added to chopped sphagnum moss, but see that the layer does not become too heavy, or it may sag badly in wet weather.

TIME TO LAYER. Spring is considered the best time to undertake air layering, usually during the month of April, but this can vary according to the season. For example, in a late season, early May or even early June may be more satisfactory. As with the majority of new techniques, some trial and error must always be allowed.

SELECTION OF LAYER. Young, clean, and healthy shoots of the previous season's growth (i.e. one-year-old wood) should be selected, usually about the thickness of a lead pencil, but here again thickness will vary according to the tree or shrub to be layered.

PREPARATION OF LAYER. First of all a longitudinal cut 3·7–5 cm (1½–2 in) long is made in the chosen shoot, in an upward direction (18a). Start the incision below a leaf-stalk or joint and continue it through the joint, with the stalk remaining on the tongue of the cut. The incision should be made about 15–30 cm (6–12 in) from the apex of the shoot. When making the cut be certain that it penetrates almost to the centre of the stem. The old Chinese method of removing a circle of bark from around the shoot has been tried, but it is not as successful as the method I have just described.

Having made the cut, treat both cut surfaces with a growth-promoting substance, preferably in powder form. There are on the market specially prepared hormone powders for air layering, which are more concentrated forms than those used for cuttings. The powder is best applied with a child's camel-hair paint brush; in this way it will be certain that all cut surfaces have an ample covering of the hormone powder. Now, see that the jaws of the cut are kept open by inserting a small wad of clean sphagnum moss between the tongue and stem (18b).

If a tubular polythene sleeve is being used, this should now be passed over the shoot and made perfectly airtight, 7·5–10 cm (3–4 in) below the cut, binding it around the lower edge with adhesive tape. Then take a large handful of good, clean sphagnum moss or peat and cotton wool free of all impurities, soak it in rain water and, after, wringing it out, pack it very carefully in the tube, or the moss or peat can be placed around the layer before covering with the sleeve (18c). Finally, the second end is sealed with the adhesive tape (18d); make sure that the tape covers each end of the film around the shoot, so preventing rain running down the shoot into the moss. This is most important. If sheet polythene is used, it should be cut into pieces about 25 × 20 cm (10 × 8 in) and wrapped around the moss so that the overlap of the film is on the underside of

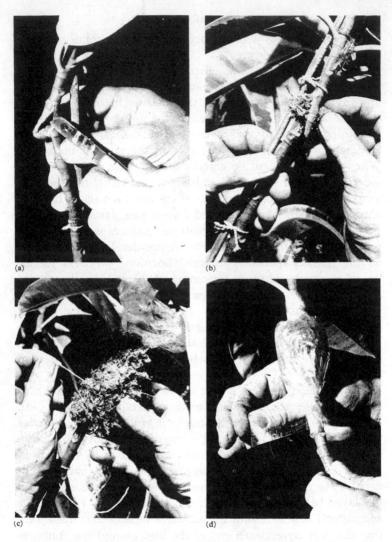

18. (a) Air layering a rubber plant (*Ficus elastica*): a cut is made through a joint to form a tongue. Or a ring of bark can be removed.

(b) The tongue is dusted with a hormone plant-growth substance and the tongue kept open with a wad of moist sphagnum moss.

(c) The treated tongue is then completely covered by sphagnum moss.

(d) Finally the air layer is covered by polythene sheeting or a sleeve and secured top and bottom by self-adhesive tape.

the air layer to prevent moisture from entering the package.

SUPPORTING THE LAYER. It is sometimes necessary to support the layer with a cane or stick in the case of thin or whippy shoots. The support can be tied to a nearby branch or to a stake in the ground.

REMOVAL AND POTTING OF LAYER. This depends largely on the type of plant and the season; as a rule, eight to twelve weeks will be needed before the layer is ready for potting. As soon as it is considered sufficiently rooted, sever the shoot by cutting off the layer beneath the lower end of the polythene package. Having removed the layer, cut or undo the adhesive tape and carefully free the roots from the moss, and before potting remove the snag (the piece of shoot beneath the roots) with a pair of secateurs. The rooted layer is then potted in a 7·5- or 10-cm (3- or 4-in) pot. Use John Innes No 1 potting compost and vermiculite in equal parts, or a soilless compost. Place the potted layer in a closed frame, preferably in a greenhouse, or in a cold frame when it should be heavily shaded. Here it can remain for about two weeks. Do not let the young layer dry out. Syringeing overhead will be beneficial, but do not overwater in the early stages, before the roots are well established.

❀ 4. Cuttings ❀

What is a cutting? Broadly speaking, it is any portion of stem, leaf or root, separated from a plant but prepared and treated in such a manner that it can grow into a new plant. The cutting differs from the division in that the latter is parted with a portion of its parent's roots and stems or leaves, and so has no need to form any entirely new organs. Similarly, a layer is made to form roots before it is severed from its parents. But in the case of a cutting, the portion that is detached and prepared is incomplete. If it is a stem or leaf cutting, it is devoid of roots, while if it is a root cutting, it is lacking stems and leaves. These deficiencies must be made good without the help of the parent plant.

EQUIPMENT AND MATERIALS

Good tools and other accessories play so essential a part in the successful handling of cuttings that I think it is best to deal with them separately before passing on to actual methods of preparation and handling.

Knives. First and foremost, the chief tool required by a propagator making cuttings is a good, sharp, straight-bladed knife. A budding knife is generally used in nurseries, but other knives can be used equally successfully, provided they are kept really keen edged. I have used on many occasions an ordinary pocket knife. For soft cuttings of such plants as dahlias, penstemons, chrysanthemums, begonias, etc., an old razor blade fixed in a piece of wood as a handle makes an efficient tool, though I prefer a knife.

66

Secateurs. Another tool required will be a good pair of secateurs that will make a clean cut.

Spade. Where cuttings are to be taken of roses, privet, bush fruits, and flowering shrubs of the larger kinds, a partly worn spade is a useful tool for digging out the slits or narrow trenches in which the cuttings will be inserted.

Dibbers and patters. For small cuttings dibbers are used, ranging in size from that of a thin pencil up to a thickness of 1·2 cm ($\frac{1}{2}$ in). To make these dibbers choose a hard wood – oak, teak or yew. For the thicker dibbers, the pegs used by carpenters in the posts of a board fence are excellent. I still use a dibber I have had for over forty years now, made out of teak. Patters are also useful for levelling and firming the soil in pot, pan, box or frame, and they may be round, square or rectangular (4). The bottom of the pot can be used with advantage in place of a patter.

Watering cans, etc. It will be necessary to have a watering can with a fine rose, but also have a spare coarser rose, for sometimes it will be necessary to water the cutting bed a while before cuttings are to be inserted. A good syringe is a most useful and necessary tool: have the type that has a turned-up nozzle and with several different-sized nozzles attached. Today there are various-shaped pump-style sprayers and squeeze types holding from 0·28–1·1 l ($\frac{1}{2}$–2 pt), and these are excellent for giving cuttings a thin mist of moisture.

Containers. Pots, pans and boxes in varying sizes are all required; also a store of large and small pieces of broken crocks, as bad drainage can be disastrous to any cutting. Earthenware pots are becoming more and more scarce, while plastic pots and polystyrene seed or cutting trays are now available.

Greenhouses and frames. A whole chapter could be written on this section. There is no doubt that those who possess glass can propagate successfully a much greater variety of plants, and still more so where heat is available. The present day advance in the electrical heating of greenhouses and frames has made it much easier for many amateurs to have this facility, when all they need to do is to set up an electrically heated propagator on a bench and plug it into an electric point, and all is then ready to take the pots or boxes of cuttings. Nevertheless, much success can be achieved where no heat is available, by using a cold greenhouse or frame. A sun frame is an ordinary cold frame, filled with sand instead of compost. It is often known as a Paris frame, and in the hands of a skilled propagator is immensely useful in the rooting of many otherwise difficult cuttings. Obviously a sun frame will require much more frequent watering than one filled with a mixture containing soil or peat. This is often a drawback for the amateur who is away from home all day. However, recent developments such as mist propagators (pp. 97–8) has enabled gardeners to have far less worries over keeping cuttings moist.

Propagators. Tremendous advances have been made in equipment in which seeds can be germinated and cuttings rooted since the second revised edition (1963) of *Simple Propagation* was published. The old French type of cloche (better known as a bell glass), as used for forcing early vegetables, has been superseded by crystal-clear polystyrene and plastic propagators. These propagators have the added advantage of fitting neatly over polystyrene and plastic trays or seed boxes, which make handy units for seeds or cuttings. They also have adjustable ventilation, while the smallest have air holes. For the amateur who has neither frame nor greenhouse, there are those propagators for use in the house which also have a drip tray. Where only a small quantity of cuttings need to be rooted or seeds sown there are pot-shaped crystal-clear polystyrene covers which fit over 3, $3\frac{1}{2}$, $4\frac{1}{2}$ and 5 in size polystyrene, plastic or clay pots.

There are, however, more sophisticated and much larger

19. The Camplex heated base is constructed throughout with PVC and rests on wooden feet. The insulated element will give a temperature rise of about 6°C (20°F). It is specially designed to take both Stewart's and Ward's propagator domes. Size of base 60 × 33 × 5 cm (2 ft × 13 in × 2 in).

propagators on the market which are like miniature green-houses which have bases made of fibre-glass, polyethylene or wood and metal. The larger designs are glazed in glass having sliding panels and a glass roof. Others have a clear plastic roof.

Then there is the type with a polyethylene base which can take two standard size and two small seed trays each having a clear plastic cover.

There are also the one tray or seed box propagators which have an electrically heated panel, beneath a clear plastic or polystyrene cover (19). There are three main ways in which these propagators are electrically heated. Those with the electric heating element bonded into the base, heating cables immersed in sand and the heating cable placed beneath slatted wood which enables the heat to circulate immediately beneath the seed trays or pots. Besides the heating element there are those that are thermostatically controlled. Manufacturers' and suppliers' addresses can be found on page 239.

Polythene stretched over a frame or box 20–23 cm (8–9 in) deep makes a useful and successful propagator. Alternatively, a sheet of glass can be used. Or pots of cuttings can be placed in polythene bags tied securely at the top. Large 2-lb jam jars (when you can find them), turned upside down over a few cuttings, act like miniature bell glasses. I well remember rooting my first aubrieta cuttings in an old box covered with a pane of glass at my home. And I did not have any well-prepared compost, just some sand and mother earth, and how sour the soil looked after the little frame had been shut up for some time, but the cuttings rooted eventually. Had I known what compost to use, and that aubrietas were better in the open, how much quicker would have been the result!

Compost. The fundamental ingredient in the compost for rooting cuttings is pure sand, but when inserted in sand alone, as in the Paris frame (now superseded by mist propagation), or propagating frame in a greenhouse, cuttings must be attended to as soon as rooted, as there is no nourishment in sand. If it is desirable to leave the cuttings in after root growth has started, other ingredients must be in the compost. For the amateur, a mixture of sand, loam and peat moss litter, or sand and peat moss alone, is generally best – or use one of the soilless composts. Sand and osmunda fibre may be used for the propagating frame in a heated greenhouse.

For general purposes, use the cutting compost recommended by the John Innes Horticultural Institute. The following has been proved successful with a great many greenhouse, hardy and bedding plants: 1 part medium loam, 2 parts granulated peat and 3 parts coarse silver sand (all parts are by volume). No fertilizers are required.

Soilless composts. When good loam can be obtained J.I. compost is first rate, but as loam has become scarcer research has been carried out into composts which do not contain it. There are now several proprietary brands of soilless composts, all based on peat, on the market, which I give in alphabetical sequence: Arthur Bowers, Baby Bio, Croxden, Kerimure, Levington and Vermipeat. No doubt there are and will be others.

The advantages of peat over loam are several: it does not usually contain any weed seeds; its decomposition is slower than soil, and it is practically sterile and therefore it is more likely to be free of disease organisms. It also has the advantage of holding water as well as air, which means it does not dry out so quickly yet does not become easily waterlogged. Its nutrient content is comparatively low so it is necessary (and easy) to adjust this by adding fertilizers. The last of the proprietary soilless composts I have mentioned, Vermipeat, incorporates vermiculite which contains up to 8 % of potassium and magnesium, plus essential trace elements. The other five soilless composts all contain a percentage of sand. All should be properly moistened, preferably before filling the pots, pans, boxes, etc.

The composts made from loam, peat or leafmould, and sand such as the J.I. composts which have been in general use for the past forty years, are still excellent and used by many amateurs and professionals alike. However, as loam becomes more and more difficult to obtain, the soilless composts are also being used, both commercially and by amateurs.

Many propagators today have as much success for the majority of plants with the soilless composts as they do with the more conventional ones. Nevertheless, it is worth remembering that there are certain plants that are better propagated in a soilless compost.

Good drainage is one of the secrets of success, whether the cuttings are in the open, frame, pot, box or greenhouse. Many hundreds of cuttings are lost by damping-off – for advice on how to prevent (see pp. 29–30).

It is also most important to have a reasonably firm cutting bed, as in the case of the seed bed (p. 24).

THE VARIOUS FORMS OF CUTTINGS

It will be most convenient to consider the selection and treatment of cuttings under certain headings, according to the type of growth from which they are prepared.

(a) **Top cuttings.** All above-ground cuttings could be termed top cuttings, but when the experienced propagator uses this term he applies it to cuttings taken from the stem itself, either the top of a main stem or a sideshoot or branch. These cuttings can be either soft-wooded, half-ripe or hard-wooded.

These terms have reference to the condition of the growth from which the cuttings are prepared, soft-wood cuttings being made from the young shoots while they are still soft and sappy, and half-ripe cuttings from the stems of plants well advanced in their season of growth and already becoming fairly firm, while hard-wood cuttings are taken from trees or shrubs at the end of the growing season when the shoots have become fully ripe and firm.

Under the general heading 'top cuttings' may also be included basal cuttings. These are prepared from shoots thrown up from the base of the plant, sometimes directly from the roots. A good example is the chrysanthemum; in general it is agreed by growers that chrysanthemums are better grown from basal cuttings than those taken from the old stem. Other examples of plants rooted from basal cuttings are lupins, delphiniums and dahlias.

(b) **Pipings.** These are really top cuttings of a special kind but they differ from most other top cuttings in that they are not prepared in any way, but are just the tips of shoots pulled out by the thumb and finger. Carnations and pinks yield cuttings of this kind very easily.

(c) **Bud cuttings.** These are a form of leaf cutting but different from the leaf cutting as dealt with a little further on. A bud cutting consists of a half-ripened stem with leaf attached, and the normal growth bud in the axil (**20c, d, e**). The cutting is then inserted in the same way as a leaf cutting. This is an effective method of propagating *Camellia japonica*, dracaenas and *Ficus elastica* (the India Rubber plant).

(d) **Eyes.** This form of cutting is closely akin to a bud cutting, and by some it may be considered 'hair splitting' to make any distinctions, but my reason for treating it separately is that it is a useful method of increasing the vine. 'Eyes' are taken in autumn or winter while the vine is dormant, and hence no leaf is attached as was the case with bud cuttings. Wisteria can also be propagated from 'eyes'.

(e) **Leaf cuttings.** This is a simple method of propagation, particularly for begonias, gloxinias, ramonda, streptocarpus and echeveria, all of which are capable of producing roots from their leaves.

(f) **Scales.** This is a simple form of vegetative propagation applied mainly to lilies. All that is entailed is the removal of the broad fleshy scales, which are then laid in moist sand or peat and in time form new bulbils and shoots.

(g) **Root cuttings.** These are portions of a root cut up into pieces, usually anything from 2·5–5 cm (1–2 in) long. *Morisia monanthos* (syn. *M. hypogaea*), phlox, seakale, romneya, verbascum, hollyhock, bouvardia, aralia and the perennial statice (*Limonium*), can all be increased this way, but by no means all plants can be propagated from portions of the root. It is interesting that the wild cherry can be grown by this method.

73

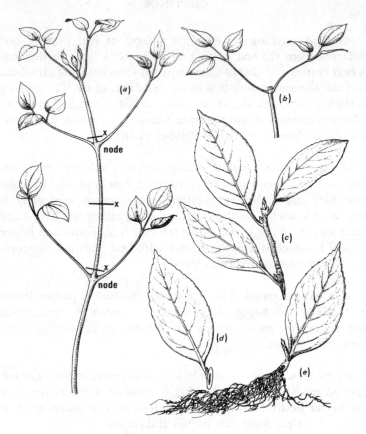

20. Internodal cutting: this is made by severing a shoot (a) between two nodes X, and later removing the pieces of shoot above each node X. A prepared internodal cutting (b) of clematis.

Bud cutting: a shoot of camellia (c) before having bud cutting removed. After removal a small portion of the shoot (d) and leaf remain attached. (e) A well-rooted bud cutting.

SELECTION AND PREPARATION

The selection of the right type of cutting is of paramount importance. It is necessary that the propagator should fully understand not only the form of cutting he wishes to take, but what type of growth it should be prepared from, particularly in regard to the degree of ripeness of the shoot.

In a broad sense there is no close season for cuttings, but for general purposes top cuttings are prepared and inserted during spring, summer and autumn, the type of cuttings taken off in the three seasons falling into the three groups, soft-wood in spring, half-ripe in summer, and hard-wood in autumn. Root cuttings are, as a rule, taken in winter.

It is essential that cuttings should be taken from plants with a strong and healthy constitution and free from disease. It will be obvious that weakly cuttings are not going to survive being severed from the parent so well as those that are strong and healthy. After preparation and insertion cuttings often have to stand up to all kinds of conditions, particularly when handled by the amateur, who will quite likely have to improvise his cutting frame and equipment. His ultimate success from these improvisations is all the more commendable.

Many cuttings are killed by insufficient and wrong preparation, although they may be prepared from the correct type of material, and taken off at the right time. I cannot impress on the reader too strongly the importance of careful preparation. Practice and experience will be required before the best results are obtained.

In the first place, the time and type of cutting to be selected must be considered. This will have a marked bearing on how the cutting should be prepared. Stem cuttings, for example, of trees, shrubs and herbaceous plants, will be cut immediately below a node or joint, i.e. a node cutting, with perhaps the exception of the clematis and shrubby and alpine verbenas, which root readily from internodal cuttings; these are cuttings severed between two nodes or leaf joints (20a, b).

The portion of cutting to be inserted beneath the soil should have its leaves and leaf stalks removed, this being done as close to the cutting as possible, without a damaging wound. Some cuttings that have buds or eyes are liable to grow and produce fresh shoots or suckers from below ground. Generally these buds or eyes should either be cut or rubbed off, but only from the portion placed underground. Blackcurrant buds are not removed in this way.

Special points. When preparing cuttings of woolly-leaved plants, it will be found an advantage to remove some of the leaves immediately above the soil level after insertion, to prevent damping-off and rotting.

Conifer cuttings. These are not of the easiest to strike on account of the resinous nature of the wood. This resin congeals at the base of the cutting and prevents the callus from forming. To overcome this, do not scrape off the resin, but dip each cutting in fairly hot water for a few minutes before inserting it in the normal way. This is not required with all conifers, e.g. *Taxus* (yew), *Chamaecyparis* (cypress), *Juniperus*, *Thuja* and others, as the resinous content of their stems is not so great as that of *Abies* (fir) and *Picea* (spruce), which are best increased by seeds or grafting, although the amateur propagator who is an experimentalist may like to try his hand at rooting these more difficult subjects.

Soft-wood cuttings. These can be taken in spring more readily than they could be in later months of the year, and require a temperature of at least 13°C (55°F). In this and the following group it is most necessary to understand the degree of shoot ripeness which will give the best result. The shoot must not be too short, too thin or too thick, for if too short the cutting will be very difficult to insert, if too thin it will in all probability not have the required strength to stand up to the check it is being given by being cut off from its parent, and if

76

21. Removing a soft-wood cutting of poinsettia (*Euphorbia pulcherrima*).

too thick it will flag readily and need much water and shading to keep it alive. For example, it would be a waste of time to insert a chrysanthemum cutting 1·2 cm ($\frac{1}{2}$ in) long, and similarly a cutting 10–12·5 cm (4–5 in) long would be too mature, and consequently take too long to root. It is most difficult to generalize as to what is the correct length for a soft-wood cutting, but approximately 5–6 cm (2–2$\frac{1}{4}$ in) long is a good average. It is equally difficult to describe the ideal thickness for a soft-wood cutting (21), but on average it is not much thinner than the lead of a pencil (except in the case of rock plants which are often thinner), and not thicker than the pencil itself. Soft-wood cuttings are usually rooted in heat.

Half-ripe cuttings. As these are taken off in summer, they are riper and firmer than soft-wood cuttings of the spring months, and are therefore rooted in cooler conditions. However, they must not be so hard that they will take a long time to

77

(a)   (b)

22. (a) Preparing a half-ripe cutting of pelargonium.
    (b) Inserting the cuttings.

form roots. As a general rule, the length of these cuttings should be about 5–10 cm (2–4 in), except in the case of heathers which are 2·5–5 cm (1–2 in) long. Good examples are bedding plants such as pelargonium (geranium) (**22a, b**), most of which can be rooted in a cold frame or with a minimum amount of heat.

Hard-wood cuttings. Many hardy shrubs, roses – especially shrub and old-fashioned roses – and bush fruits can be increased from cuttings of this type. Such cuttings are taken in autumn and winter from the current year's growth as a general rule, sometimes with a heel (**24**) but mostly without. For general guidance, cuttings should be 23–30 cm (9–12 in) in length (**23a**). It is necessary that the cutting should be fully mature and plump, as it has to stand up to more trying conditions than the soft-wood or half-ripe cutting. If on the immature side, it would in all probability die before roots had been formed. Quite often hard-wood cuttings will remain dormant in the ground all through the winter, even though they may

(a)

(b)

23. (a) Hard-wood cuttings of oval privet (*Ligustrum ovalifolium*) (right) and prepared (left) ready for insertion.

(b) The cuttings being inserted in a narrow trench to which sharp sand has been added.

24. Removal of a heel cutting from a rose shoot.

have formed a callus in early or late autumn, but as spring approaches new roots will be emitted, and signs of growth will also be evident above ground.

Occasionally, when hard-wooded subjects are required urgently for some purpose, the plants can be made to produce soft-wood or half-ripe cuttings by placing the stock plants (those from which the cuttings are to be taken) in heat.

If taken at a joint, side-shoots require the same treatment as just prescribed, but if taken off with a portion of the main stem or branch, they are termed heel cuttings (24) and the small, heel-like portion attached will be a little ragged and must be pared off clean with a sharp knife. Some gardeners think that these heel cuttings root more reliably than cuttings severed beneath a node, but my own opinion is that there is little if any advantage to be gained and that sometimes the heel cutting is actually slower in rooting. Other preparation, such as removal of leaves and buds (where necessary) is carried out in the usual way for the portion inserted in the soil. Where cuttings of large-leaved subjects are taken, it is often necessary and wise to cut off a portion of the leaves at the top of the cutting to reduce transpiration.

Vine eyes. Material wanted for vine eyes is taken from prunings, in November and December, from well-ripened laterals; tie them in bundles, label them and heel them in to half their length out-of-doors. In mid-February, select those

25. Vine eyes being prepared from shoots 3–4 cm (1–1½ in) long with a slanting cut made on the opposite side to the eye or bud.

26. Prepared vine eyes inserted in turves and securely pegged either end of shoot.

shoots that have good plump buds, cutting the shoots into 5-cm (2-in) lengths, with a plump bud in the centre of each piece of shoot **(26)**. The method is similar when applied to other plants, though of course the measurements will need to be modified according to their size. Alternatively, a piece of shoot 3–4 cm (1–1½ in) long with a slanting cut made on the opposite side to the eye can be used **(25)**.

Leaf cuttings. These are prepared from well developed leaves. Actually, very little preparation is required, though sometimes after their removal a number of incisions are made with the point of a sharp knife across a pair of veins and as near to where they join as possible, but not directly on the base of the V **(27a)**. Frequently, however, leaves may be used without this slicing or any other treatment.

This method of propagation is used for streptocarpus **(28)**, gloxinias, begonias, ramondas, and many other plants. Streptocarpus and gloxinias often have a portion of the top part of the leaf removed – this is to cut down transpiration.

Root cuttings. There is as little preparation required for root cuttings as there is for leaf cuttings. The main point is that a root cutting is cut horizontally at the top and slanting at the base, which is trimmed off at the tip, as this slight trimming prevents it becoming damaged during insertion. The chief reason for making the cut on a slant at the base is to prevent cuttings from being inserted upside down. Even this is relatively unimportant, for certain forms of root cuttings such as phlox may be laid on the soil horizontally. Root cuttings should be anything from 2½ cm (1 in) up to 5 cm (2 in) long. Another way to prepare these cuttings is to slice off the top growth, and the roots that remain in the ground will callus and emit new shoots.

Scales. These provide a ready and easy method of propagating lilies. All the preparation entailed is to fork up a bulb

(a)

(b)

27. (a) A leaf cutting of *Begonia rex*. Cuts are made in veins on the back of the leaf.

(b) A *Begonia rex* leaf showing young plantlets on top of leaf.

carefully and remove the fleshy scale-like leaves, which form the bulb. If the scales are fresh and plump they will come away very easily, and if not they may be assisted by a slight leverage with the point of a sharp knife. If the bulb is rare or the owner cannot afford to sacrifice the complete bulb, a few scales may be taken off all round by just scraping back a little soil and leaving the bulb *in situ*.

Lily scales may be removed and inserted at almost any time of the year, but for preference it is better to take them off after the growth has ripened (September), for by then the bulbs will have obtained all the necessary nourishment from the foliage.

PLANT GROWTH SUBSTANCES

Today the scientist plays a great part in horticulture and in recent years many experiments have been carried out by the scientist and the practical gardener with plant growth sub-stances to hasten the rooting of cuttings. Charles Darwin was perhaps the first English scientist to realize the possible existence of growth-regulating substances among plants. During the last thirty years or so, many experiments have been carried out by scientists in the laboratory and practical gardeners in the propagating house, with the result that certain cuttings, at one time considered difficult or almost unrootable, can now be rooted with ease by the aid of root-promoting hormones known also as growth substances.

There are on the market today several root-promoting substances in powder form, under various proprietary names such as Seradix and Strike (both May & Baker products), and Murphy's. Seradix comes in three grades, No 1 for soft-wood cuttings, No 2 for semi-hard-wood, i.e. half-ripe cuttings, and No 3 for hard-wood cuttings. Strike is an all-purpose rooting powder with fungicidal chemical captan added. When using Strike, dip hard-wood cuttings in water before dipping them in the powder, and with soft-wood and half-ripe cuttings, dip

1·2 cm ($\frac{1}{2}$ in) of the stem at the base of each cutting in the powder, then shake off any excess. Murphy's hormone rooting powder also contains captan, as well as naphthylacetic acid: with soft-wood and half-ripe cuttings, dip 1·2 cm (1 in) at the base of each cutting in the powder, then tap it on the side of the canister to remove any surplus; with hard-wood cuttings, dip them first in water and then in the powder, and after shaking off any surplus powder, insert the cuttings.

Provided these powders are used strictly in accordance with the manufacturers' instructions, excellent results may be obtained.

METHODS OF INSERTION

In the first place it must be determined what form of cutting is to be inserted, whether it is to have greenhouse, frame, or plastic or polystyrene propagator treatment, with heat or without, or is to be placed in the open without any protection whatsoever. The kind of compost required for the cuttings must be considered. Do they root best in sand, peat moss litter, or perhaps a particular mixture specially prepared for rooting cuttings as described on p. 70, or in one of the soilless composts? Moreover, cleanliness must be a watchword throughout, as dirty frames, pots, pans or boxes and sour composts can have serious effects and soon play havoc with a batch of cuttings.

Provided all these matters are taken into account, it is quite possible to get 90 per cent and often nearly 100 per cent strike of most cuttings.

In these advanced and constantly changing times, with greater knowledge of gardening in general, it is perhaps difficult to lay down hard and fast rules. But if the following principles are observed, there is no reason why success with cuttings should not be achieved.

Insertion of cuttings under glass, in pots, pans, boxes. For soft-wood cuttings a very light, sandy compost is required, as it is desirable to root the cuttings quickly. For half-ripe cuttings, the mixture can be slightly less sandy, as rooting will take a little longer and therefore more nourishment is required. Hard-wood cuttings have to stand for a long time, so a more bulky mixture is preferred.

Where pots, pans or boxes are used, or for that matter even when cuttings are rooted directly in beds of soil made up in frames, drainage must not be overlooked. In the case of the containers, a third of the total depth should be filled with some form of drainage, such as a good stopper (large crock) over the drainage hole, followed by smaller pieces of crock (broken pots) or cinders, and finally a cover of dry leaves or peat riddlings. This last is to prevent the soil falling and clogging the crocks. Soil should be placed to within 0·6 cm ($\frac{1}{4}$ in) of the rim or top. When soilless composts are used, the crocking of pots or boxes is not necessary.

Firmness of soil is also important. It must not be too firm, neither must it be spongy. Put the soil in with the hands and press gently with the fingers, finally finishing off with the bottom of a pot as presser or patter, that is if you have not a patter as illustrated (4). On top of the soil sprinkle a little coarse dry sand, the reason for this being that when the cuttings are inserted the sand will trickle down into the holes and aid rooting. Soilless composts are not firmed like those that contain soil: all that is required is to firm gently, but very lightly, and shake the pot or box on a bench, then give it a thorough watering, and after draining the cuttings can be inserted.

Some gardeners favour insertion of cuttings by pushing them into the soil until firm. I prefer a dibber and think it best to have several in various sizes, though for general purposes one the thickness of a fountain pen is quite satisfactory. Now if the compost is of the right texture and nicely firmed, the dibber should go in with comparative ease, and the sand will trickle down for the base of the cutting to rest on. But as a foreman of mine used to tell me; 'don't 'ang 'em'; he was right, for the base of every cutting must rest firmly on the soil.

The cutting inserted, the dibber can be used to press soil gently to its side to fill in any space left, but the utmost care must be taken not to injure the base of the cutting. Follow this by a watering with a fine-rosed can, which further settles the soil around it. In many instances the watering will be sufficient to firm the cuttings.

Insertion of cuttings in the double pot. It has been found that certain cuttings (e.g. the Cape heaths) root more readily if inserted close against the pot or pan and for this purpose a smaller pot or pan is placed inside a larger size. The inner one can be inserted in the normal position, or upside down. Both pots are crocked in the normal way if the inner one is placed upright and the inner pot or pan is filled with sand. If inverted, it is filled with sphagnum moss. The outer one is filled with a sandy cutting compost. Cuttings are then inserted between the outer and inner pots. The advantage of this method is that it reduces the volume of soil, and ensures perfect drainage and aeration.

Insertion of cuttings in the open. Some cuttings are inserted either in the open frame (without glass) or in the open garden, though a specially prepared cutting or nursery bed is more desirable. Hard-wood cuttings are usually inserted in the open. Smaller cuttings such as those of alpines and herbaceous plants, bedding plants and some shrubs, are inserted in a frame without the light, though a handlight may be placed over the cuttings for a few days and then removed entirely.

Cuttings that require the slight protection of a frame are best dibbled in, followed by a good watering. Larger hard-wood cuttings, for example roses, flowering shrubs, bush fruits and fruit stocks, may be inserted by one of three methods as follows: (a) by dibbers; (b) by making a V-shaped slit in the ground with a spade (a partly worn one is best), but care must be taken to ensure that the base of the cutting is touching the bottom of the slit (see p. 79) and not suspended with a cavity beneath – after the cuttings are inserted, they should be firmed by treading with the feet; (c) the last method is to dig out a narrow trench deep enough to insert the cuttings to two-thirds their length (23b). On the bottom of the trench sprinkle some coarse sand, as this encourages good rooting and also creates better drainage at the base of the cutting. The cuttings are then inserted at the back of the trench. By this method there is no fear of the base of the cutting not resting on the soil, which is essential. The cuttings inserted, fill in with soil, treading them in firmly as prescribed for cuttings in the V-slit.

Insertion of bud cuttings. These are inserted in the same way as other cuttings and as rooting takes place the young buds will grow away to form the new plants. This operation can be carried out with or without heat, according to the plant being propagated.

Insertion of vine eyes. There are two methods of inserting vine eyes, i.e. by using turves or by using pots. Where turf is used, fill a seed tray size $38 \times 23 \times 7.5$ cm ($15 \times 9 \times 3$ in) with four pieces of turf placed turf side down, each about 15 cm (6 in) long, 7.5 cm (3 in) wide and deep. In the centre of each section make an indentation in the turf, filling it with sand and pushing in the piece of shoot with the eye, so that the bud stands up above soil level (26). Secure the shoot each end with a layering pin. If using 7.5-cm (3-in) pots fill them with a mixture of loam and sharp sand, inserting one eye cutting in each pot. In either case, after insertion, water the pots or turves. Little or no water should be needed until new growth starts and, even then, go easily with the watering can.

Pots or trays of turves should be placed in a propagator or greenhouse frame where there is bottom heat to a temperature of 16–21°C (60–70°F). Greenhouses should have a temperature of 13–16°C (55–60°F). Rooting will take about five or six weeks to occur.

Finally, pot into 12.5-cm (5-in) pots in a compost of fibrous loam, charcoal, sand, bonemeal and burnt earth. Do not use leafmould.

Insertion of leaf cuttings. There are two methods of inserting leaf cuttings. One is to lay the leaf down flat, having previously made a number of cuts in the midribs and other ribs (27a, b). The leaf is laid flat on the compost consisting of sand and peat moss. At one time it was the practice to peg the

28. Top right: *streptocarpus* leaf prior to preparation. Top left: the rooted leaf, the top of which has been removed to save loss of moisture. Below: the rooted cutting potted.

(a)

(b)

29. (a) Lily scales of *Lilium Martagon* removed from older bulb, being inserted with the aid of a label.

(b) Lily bulbs produced from scales, a young bulb can be seen at the base of a scale.

leaves down, but it has been found better to do without pegs as these increase the chance of the leaf rotting.

The other method is to insert the leaves upright or flat – ramondas are inserted in this way. With streptocarpus (28), for example, a portion of the top half of the leaf is removed to cut down transpiration. Gloxinias can also be inserted in an upright position, the stalk of the leaf being fixed in the compost around the edge of a pot.

The propagation of leaf cuttings is usually carried out in a warm greenhouse, with a temperature of 16–18°C (60–65°F).

Insertion of scales. Lily and other bulb scales are stood upright just like cuttings, with their tips standing out above the soil (29a) and inserted in boxes, pans or pots containing a compost of 2 parts medium loam, 1 part oak or beech leafmould and 1 part coarse, sharp sand. To this add a good sprinkling of charcoal. If the propagator wishes to leave the bulbs until good bulblets have been formed, he should not fill the pots, pans or boxes too full but leave a space of 2·5–3·7 cm (1–1½ in), to allow for further filling up with compost as the scales turn into new bulbs (29b). The best time to carry out this operation is usually as soon as the bulbs have finished flowering.

Insertion of root cuttings. Propagation by root cuttings has always interested me; whether this is because it is such a simple method I do not know, but insertion and cultivation are indeed easy once the root cutting has been prepared as described on p. 82. All that is required are pots, pans or boxes filled with a compost of loam, leafmould or granulated peat, and sand. The root cuttings of most plants can then be dibbled in in the normal way with a dibber adapted to the size of the root cutting. It is as well to have a dibber with a blunt end rather than pointed. The cuttings must be kept right way up when this method is used and should be sunk just below soil level. There is another method of insertion, namely by placing the roots on top of the soil and covering with a further

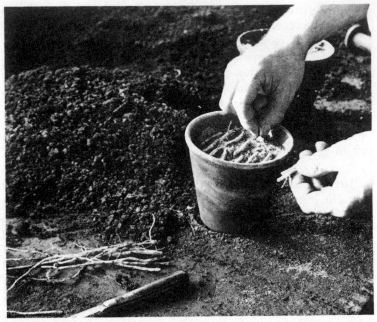

30. Pieces of root cuttings of *Primula denticulata* are laid horizontally in a pot filled with compost and then covered lightly.

layer. The cuttings are then, of course, on their sides, but this does not matter. This method is used for phlox (p. 82), *Primula denticulata* (**30**), and *Anemone × hybrida*, chiefly because their roots are thinner. As a general rule, root cuttings can be rooted in an unheated greenhouse or frame, though sufficient heat to keep frost out is naturally an encouragement to quicker growth. Root cuttings are inserted during autumn, winter and spring, but principally in winter.

AFTER-CARE OF CUTTINGS

Success with cuttings depends to a very large extent on the care and attention given during the time roots are forming.

It is essential that adequate moisture is maintained both in soil and air, for cuttings should never be allowed to wilt, and this is of particular importance in the case of cuttings inserted in sand alone. Correct temperature and an adequate supply of light are also important and aeration of the soil in which the roots are to develop is most vital; that is why careful attention must be paid to good drainage.

Generally speaking, soft-wood and half-ripe cuttings root more readily when given a close, moist atmosphere, and to maintain this type of environment frequent syringeing and little or no ventilation are necessary. Conditions such as these induce quicker rooting, because the flow of sap is stimulated. Hard-wood cuttings prefer a cooler and freer atmosphere and even when in a frame may need some ventilation. All cuttings should be kept in a fairly moist state, whether inserted in frame or greenhouse, under a plastic propagator, or outside without any form of protection, for should young roots have formed and, just at the critical stage, the cutting medium be allowed to become dry, the roots would wither and the nourishment remaining in the cuttings would, in all probability, be insufficient to form fresh roots. On the other hand, waterlogged soil is equally fatal to cuttings.

Certain cuttings require an abundance of moisture, yet cannot stand a very wet condition at the point where roots are to form. This difficulty can be met by using the double pot method described on p. 87.

Where summer cuttings are inserted in sand alone, without artificial heat, but in a frame with the light (i.e. the glass covering) kept closed until cuttings are rooted, one good watering after insertion of cuttings should be given and further waterings whenever the sand becomes dry on the surface; shading will be necessary until they are rooted.

Frames covered with glass or plastic propagators outside will not usually require more than one watering after insertion, but it is as well to look at them once a week, in case they are in need of water. In very hot weather transpiration can be reduced by a slight stippling of the glass with whitewash or covering with a fine muslin (tiffany).

Cuttings inserted in propagating cases or plastic propagators inside a greenhouse must be examined daily, and the light or propagator should be wiped free of condensation night and morning. Then water with a very fine-rosed can or a syringe and close down at once to create the necessary moist atmosphere. If shading is needed, use newspaper or thin brown paper, as this is easily removed when not wanted.

Hard-wood cuttings that are inserted either in a frame, open or otherwise, or in the open ground, as a rule only require an initial watering immediately after insertion. Even those under a light will not require more, because hard-wood cuttings are made in the autumn when the sap is inactive, sun power is less and days are shorter, and therefore the rate of transpiration is considerably reduced. After a severe frost, cutting beds should be gone over to see whether the frost has raised the cuttings up out of the ground. This will often be the case and if left in this state the cuttings will in all probability be killed. To remedy this go over the beds and push all raised cuttings back into position, then press the soil back and eventually firm it by treading with the heels. It is in the spring of the year, when these hard-wood cuttings begin to form roots, buds and foliage, that moisture will be called for, also a slight stirring of the soil and the removal of weeds as they appear.

Although the temperature required by any individual or group of cuttings varies, a general guide can be given. Soft-wood cuttings, whether stem or leaf, require more heat than do half-ripe or hard-wood cuttings. Such soft-wood cuttings inserted in a closed propagating frame, or propagator or under mist – all housed in a greenhouse – require on average a temperature of 16–18°C (60–65°F). For plants of a tropical nature, a temperature of 21–29°C (70–85°F) is necessary.

Half-ripe cuttings will root readily in a temperature of 13–16°C (55–60°F).

Hard-wood cuttings are kept quite cool.

In deciding what the temperature should be for any given plant, it is also necessary to take into consideration the conditions under which the plant has been growing before cuttings were removed. For example, dahlias, although grown outside as hardy plants during the summer, have to be brought into a greenhouse in spring and forced into growth in heat, and the cuttings are, in consequence, inserted in a slightly higher temperature to induce quick rooting. This is necessary on account of the softness of the cutting.

Some plants, such as succulents, require full light all the time, but in most cases soft-wood and half-ripe cuttings need some shading at first to reduce the rate at which they lose water. However, even when shading is used, the propagator should watch his cuttings carefully and remove the shading before they become drawn or spindly. Much of this kind of knowledge must be acquired by one's own observations, but a sign that shading should be removed is the appearance of new growth, buds or small leaves. As soon as the shading is removed the leaves will become greener and stronger.

Soft-wood cuttings root quickly, in a matter of a week or so, particularly when given some bottom heat. Half-ripe cuttings take as a rule two or three weeks or even a month or so. Hardwood cuttings that are inserted in the autumn may possibly callus over the same year, but roots and shoots will not be put forth until the following year.

TREATMENT OF ROOTED CUTTINGS

No real advantage is gained by leaving cuttings in the pot, pan, box or frame once they are rooted. This applies particularly in the case of cuttings rooted in sand only, for there is no nourishment to be had from sand; so the sooner the young cuttings are potted on into a good, light, open, sandy compost as described on p. 32, the better. No strong foods in the way of manures or fertilizers are required at first, for the young plants cannot take it, any more than a baby can take beef. Try and place the cuttings in a similar temperature to that in which they were rooted, gradually taking them into cooler temperatures as they begin to grow and become accustomed to their new mode of life. At first a little shading with paper is usually required, unless the weather is dull.

With young plants even more attention should be paid to cleanliness of pots. Sizes should be used that are large enough to take the roots without cramping them, but overpotting must be avoided. For general purposes 6·2-, 7·5- and 8·7-cm (2½-, 3- and 3½-in) pots are large enough for first potting. Do not overlook good drainage. As the young plant grows and fills the pot with roots, it may be repotted into a larger size, using a little richer compost, according to the individual plant. Plants that grow rather rapidly in height or are easily broken, will require supporting by small hazel sticks or split canes.

Do not allow newly potted cuttings to flag or dry out.

All these foregoing remarks relate chiefly to cuttings rooted with some heat, such as soft-wood and half-ripe cuttings. Hard-wood cuttings benefit by gentle heat when first potted to assist root growth. Many hard-wood cuttings can be planted direct into nursery beds, particularly subjects like roses, shrubs and bush fruits. Such cuttings are not lifted from the cutting bed until a year after insertion.

MIST PROPAGATION

Mist propagation is, like air layering, only an adaptation or improvement on an earlier method of propagation. In fact for nearly half a century the 'sun-frame' method has been used, though not so much today as it was, because such a method entails frequent syringeing of the cuttings, and the present day labour situation does not make this economical. Today, however, mist propagation has made this syringeing entirely automatic. The main purpose of mist propagation is to envelop the cuttings with moisture in order to raise the relative humidity. The actual wetting of the leaves is immaterial. What

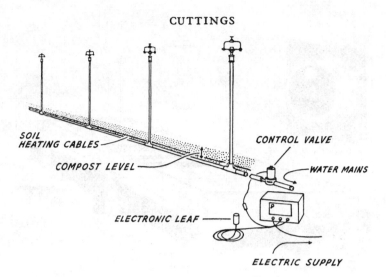

SOIL
HEATING CABLES

COMPOST LEVEL

CONTROL VALVE

WATER MAINS

ELECTRONIC LEAF

ELECTRIC SUPPLY

31. A diagrammatic drawing of a mist propagating unit.

is important is that transpiration is very considerably reduced
by the high relative humidity.

For the amateur, the most satisfactory method of mist
propagation is the use of the 'artificial' or 'electronic leaf', though
in point of fact it neither resembles a leaf nor is electronic (**31**).
What this instrument does do is detect by a controlling electrical circuit the need for moisture when it automatically turns
on the water supply. The 'leaf' is placed among the cuttings,
and so long as it is moist the water is turned off, but as soon as
it becomes dry the water is turned on for a predetermined
period and is applied through jets fixed above the cuttings to
give a fine mist. The apparatus can be used on an open bench
in the greenhouse (**32a, b**) or in a frame so that the cuttings
get full air and sunlight. Some prefer to have it working in a
polythene tent, but this is not essential. What is essential is
to have a really free mist which damps the cuttings without
saturating the cutting bed.

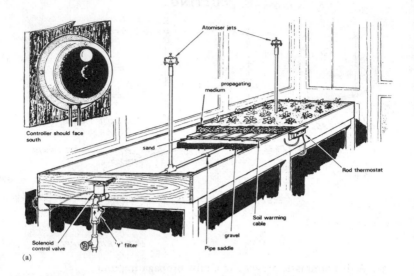

Atomiser jets

propagating medium

Controller should face south

sand

Rod thermostat

Soil warming cable

Solenoid control valve

'Y' filter

gravel

Pipe saddle

(a)

(b)

32. (a) A typical solar mist layout.
    (b) A mist propagating bench.

Cuttings. A great variety of plants can be successfully rooted under mist. It is particularly advantageous with large-leaved evergreen cuttings, where transpiration is rapid. In this respect it is very necessary to keep the cuttings turgid after collection and today the propagator has the advantage of the polythene bag in which to place his cuttings.

The length of a cutting does not appear to make any difference to its success in rooting. The most difficult to root are hairy-leaved plants, e.g. artemisias, as indeed they always have been. But in rooting any cutting by this method (or any other for that matter) the selection of suitable cutting material is still as important as it ever was.

Shade and heat. With regard to shading, it has been found beneficial to shade the cuttings during very hot, sunny weather.

It is an advantage to have bottom heat with a temperature of 21–24°C (70–75°F). To obtain the required heat an electric cable is highly satisfactory. It is, however, an advantage to have it thermostatically controlled. Low or main voltage wiring can be used, with a loading of 12 to 15 watts. Without the bottom heat, cuttings will take much longer to root.

Growth-promoting substances. These hormone growth-promoting substances are of value with the more difficult-to-root cuttings, but otherwise there is really no need to use them.

Containers. Cuttings can be inserted direct into the open bench or in pots, pans or boxes, but whether bench or container is used, good drainage is the keystone to success. Plenty of crocks must be used which must be covered by coarse leaves or peat. With boxes it is best if a 0·6-cm ($\frac{1}{4}$-in) wire mesh base is used instead of the normal wooden one.

Compost. I cannot do better than recommend the compost used at the Royal Horticultural Society's Garden, Wisley, where it has been found most successful.

For the majority of plants, whether non-ericaceous or acid-loving, use equal parts of medium granulated peat and Bedfordshire sand. Both parts by volume.

Bedfordshire sand, if you can obtain it, is ideal. Otherwise use a good, sharp, local sand or, as an alternative to sand, Silvaperl Perlite.

For small cuttings put the peat through a 6 mm ($\frac{1}{4}$ in) sieve. If the peat is dry, make sure it is thoroughly moistened before mixing with sand or Perlite.

After treatment. With very soft cuttings it is important not to let them receive a check, and as they continue to grow when making roots under mist, they must be potted and transferred immediately afterwards into a polythene nursery tent in the greenhouse before being hardened off in more open conditions.

With half-ripe cuttings this transference is not so vital, as they are less likely to make so much top growth and, therefore, can be left until their root system is more advanced. Hardwood cuttings can remain in the cutting bed until the following spring.

If any reader wishes to know more about this subject, the book to consult is *Mist Propagation and Automatic Watering* by H. J. Welch (Faber and Faber).

❧ 5. Budding and Grafting ❧

In this chapter quite a different form of vegetative propagation is dealt with, one which is used when, perhaps, all other vegetative methods have failed, e.g. division, layers or cuttings. This form of multiplication is much used by the professional gardener and nurseryman, primarily because it is quick and therefore cheap, and because it ensures uniformity in the variety of plant, tree, shrub, fruit or flower that is being multiplied.

How does this form of vegetative propagation differ from such methods as division, layers or cuttings? In essence grafting (and budding is only a form of grafting) is the art of uniting or joining two living portions of plants so that they form a permanent union, in which one (known as the stock) supplies the rooting system and the other (known as the bud or scion) supplies the branches or shoots. There are exceptions to this as to other rules, e.g. bridge grafting, which is the forming of pyramidal cages as carried out with fruit trees and the joining of trees to form a growing arch. These exceptions are not forms of increase at all and the definition given is true so far as the subject of this book is concerned.

For some reason or other the nurseryman has always been very loath to divulge the secrets of budding and grafting, apparently fearing that if he did so his livelihood might be jeopardized. In fact this is far from being the case, for the gardener who tries his hand at budding his own roses or fruit trees is usually a keen amateur, and is always on the lookout for new plants, and will usually buy when the plant is new and expensive. He is eager to try his hand at obtaining more, and in this way the new introduction becomes better known and those who have not the inclination or gift to produce their own plants, buy them from the nurseryman.

An essential that must never be overlooked is that if any budding or grafting is to be successful, the scion must be compatible with the stock, and in most cases both scion and stock must be of the same genus, though there are exceptions, e.g. a broom (*Cytisus*) can be grafted on to *Laburnum*, and lilac (*Syringa*) can be budded or grafted on to privet (*Ligustrum*). Although in these exceptions the scions and stocks are of different genera, they are in all cases the same families, i.e. *Leguminosae* in the case of the broom and laburnum, and *Oleaceae* in the case of lilac and privet.

SELECTION AND TREATMENT OF ROOTSTOCKS

The rootstocks for budding or grafting are of paramount importance, for however good your buds or scions are, failure will most certainly be frequent if wrong or poor quality stocks are used.

Rootstocks must be clean, healthy, young and vigorous – it is quite as necessary to spray apple and rose stocks as it is the already mature trees. Stocks should be uniform, true to type and well rooted. A great deal of research work has been carried out at the East Malling Research Station and at the John Innes Horticultural Institute (now at Norwich, but originally at Merton, Surrey) concerning the influence of certain fruit stocks, and the same applies to roses at the University of Manchester and other centres.

Fruit stocks. The various apple stocks go under numbers preceded by a word or letters, Malling, or M, and Malling-Merton, or MM. These stocks can be classified in groups, each with some outstanding influence on the eventual development of the tree. Such groups may be distinguished as follows: Group (a) Very Dwarf, e.g. M9 (Jaune de Metz) – stocks of this type produce a relatively small tree but one which is quick cropping; Group (b) Semi-Dwarf, e.g. M7, M26 and M106 – trees on these stocks require more room and will not bear quite so quickly; Group (c) Vigorous, e.g. M2 (Doucin) and MM111

– these stocks give larger trees than those in Group (b) but will usually start to crop about five years after budding; they are particularly suitable for large bushes and small standards or for trees grown on poor soils; Group (d) Very Vigorous, e.g. M16 and M25 for large bushes and standards. Any of these stocks or the common crab can be used satisfactorily for ornamental crab apples.

Pears are budded and grafted on various Quince stocks and selected forms are available, known as Malling Quince A (Angers), moderately vigorous, Malling Quince B (Common Quince), also moderately vigorous, and Malling Quince C, dwarfing. None of these stocks are suitable for standards, for which free pear stock is used, i.e. pears raised from seed. Commercially, much of this comes from America and is usually of the variety Bartlett, but any variety can be used.

Plum stocks are more varied. Myrobalan B (Malling Selection) is recommended for strong trees of plums only; this stock buds easily. Brompton Stock is similar in vigour and specially recommended for peaches and nectarines, which grow well on it. St Julien A is less vigorous and suitable for smaller bushes and trained trees of plums, peaches and nectarines. Pershore Plum Stock is useful for smaller trees. Common Mussel is now being largely superseded by St Julien A, though both are useful stocks for ornamentals belonging to the prunus family, e.g. *Prunus persica* and *P. dulcis* (syn. *P. amygdalus, P. communis*) – flowering peaches and almonds – and *P. cerasifera* 'Pissardii' (purple-leaved ornamental plum), which also does particularly well on Myrobalan B or seedling Myrobalan.

Cherries are budded and grafted on to rootstocks of *Prunus avium*, the common gean or wild cherry, that we see so much of in the spring in woods and hedgerows, and also Mazzard which has a very hard stem. A selection of *P. avium* F12/1 is particularly reliable for vigorous and even growth. What propagators are searching at present for is a more dwarfing rootstock for cherry.

Rose stocks. There are many types of rose stocks that can be used but not all are desirable. Each of the following has some advantage or special purpose for which it is most suited: Briar seedlings (*R. canina*); Briar cuttings (*R. canina*); Rugosa cuttings; Multiflora; Laxa; Manettii cuttings.

Briar seedlings are easily raised from seed. Hard-wood cuttings of briars can be taken from hedgerows in autumn. At the same time as the cuttings are chosen, look out for stems for budding as standards, half-standards and weeping standards. Rugosa stocks grown from cuttings have a large coarse root system and produce very large maiden bushes, but these do not last like those grown on briar stocks. Rugosa stocks are not satisfactory on chalk soils. Multiflora (also known as Polyantha Simplex) provides a fine even stock, with a nice clean neck for budding and a good fibrous root system. Laxa, correctly known as *Rosa canina* 'Froebelii', gives vigorous growth and, being nearly thornless, is easy to work. The Manettii rose grown from cuttings, makes a useful stock. It buds well and puts vigour into weak varieties, but does not always make a good or permanent union. Moreover, suckers from Manettii are extremely difficult to recognize. However, it is a useful stock for roses grown under glass.

Preparation of stocks for budding and grafting. Once the stocks are acquired it is essential that they should be well looked after. First and foremost the ground should be clean of weeds, particularly perennial weeds, well cultivated and in good heart. If the ground is in poor condition, rotted farmyard manure should be added. Manure placed in the trench when planting is the best method. Plant in autumn as early as conditions permit, for the more established the stocks are, the better will be the results. Stocks required for budding are planted in October or November and budded the following summer. Stocks required for grafting can be planted at the same time but are not grafted until the second March or April after planting, when they are fifteen to eighteen months old. Clematis are grafted on to the roots of *C. vitalba*, which are lifted in the autumn, potted up or laid in and grafted in February or March, in heat.

Care of stocks after planting. Newly planted rootstocks should be encouraged to make good healthy and clean growth by frequent cultivation throughout the spring and summer months, as the growth made during this time will determine the future growth of the young tree after budding or grafting. In the case of fruit and ornamental tree stocks, the first 23 cm (9 in) of growth buds on the main stem should be removed. If this is done when they are young, green and soft, they are easily removed by gentle rubbing with the hand.

BUDDING

Budding of fruit trees, roses and ornamental.trees and shrubs is carried out during the summer months, approximately from June to September, though in most seasons June to mid-August is considered most satisfactory. The actual operation of budding takes place when the rind parts easily from the wood of the stock, and when the buds have developed on the shoots from which they are to be taken. These buds are situated at the base of the leaf stalk.

If the weather should be very dry and the plants are so short of sap that rind clings to wood, wait until sufficient rain arrives to make the rind run easily; or it may be necessary to refrain from budding altogether and instead graft the following spring.

Selection of bud sticks. In the case of roses, the buds are best taken from shoots that have already flowered, as on these the buds will have plumped up. The best buds are those situated in the middle of the shoot. A test for the right condition at which to take the buds of roses is when the thorns snap off easily, leaving a damp green patch underneath.

When taking off shoots for budding sticks, it is most important that a label is placed on each variety. A record should also be kept of the number of buds put on of each variety. When stocks are planted in rows, it is easy to check what the variety is, even if the label is lost, by counting up the number of stocks from the start of the first row.

Removal of bud sticks. At all times the buds must be kept fresh and plump, i.e. from the time the shoot is cut off up to the time the bud is removed and inserted. The shoots should have all leaves removed, leaving only a small portion of the leaf stalk, approximately 0·6 cm ($\frac{1}{4}$ in) to 3·7 cm ($1\frac{1}{2}$ in) long. This is left to act as a handle when the bud is inserted in the stock.

Budding bush roses. Although the budding of roses, fruit trees and ornamental trees and shrubs is carried out on very similar lines, there are a few points that apply individually. For the sake of simplicity, I have taken roses as the general example.

One outstanding difference between budding a bush rose and a fruit tree is that with a rose the bud is inserted as close to the roots as possible, whereas with the fruit stock the bud is always kept well above the soil level. The reason for this is that it is an advantage if eventually the rose is more or less on its own roots, whereas the fruit stock has probably been chosen specially for the influence it will bear on growth and fruitfulness, and it is not desirable that this influence should be upset by roots of a different character produced by the scion. A good budder should always study the eventual effect.

When preparing for budding, the soil is scraped away from all round the stock. In the nursery this is done with a hoe, but for the amateur a hand trowel is an ideal tool. The soil is not put back after budding but left clear of the stock. To ensure a clean stock, a piece of damp flannel may be used to wipe it down before budding.

The stock is now ready for budding. For the insertion of the bud a T-cut is made **(33d)**. First make the cross-cut of the T, then make the stroke of the T in an upward direction to meet the cross-cut. The budding knife used should have a handle flattened at one end, like a spatula or paper knife. This portion is used to prise open the flaps of rind formed on each side of the T. If this operation cannot be carried out easily, then the stock is not ready for budding.

The bud sticks must be kept fresh and for this purpose they should be stood in a jar or small tin in 5 cm (2 in) of water. No more water should be used, for buds must not be wet when inserted in the stock.

The bud is now removed from the bud stick **(33a)** by making a slanting cut starting 1·2–2 5 cm (½– 1in) below it. It is as well to tear the bud away at the finish rather than let the knife complete the operation. The bud then remains with shield attached **(33b)**. Hold the bud in the left hand by the portion of petiole (leaf stalk) that remains, and with the right thumb and first finger take hold of the sliver of wood behind the bud and sharply lift it upwards and pull backwards towards the base of the bud with a quick bending motion, leaving the whole clean except for the little knob or heart-shaped protuberance (bud trace) in the shield **(33c)**. If instead of this knob there is a hollow, the chances are that the bud trace has been torn out and failure will usually result, though occasionally there are exceptions.

A debatable question is on which side of the stock the bud should be placed. Some say that the incision should be made on the side most protected from the sun, while others say on the side that catches the prevailing winds. Then when the wind comes, the bud is not blown out but instead kept securely in the stock. In Bulletin No 135 of the Ministry of Agriculture, Fisheries and Food, *Fruit Tree Raising, Rootstocks and Propagation*, published in 1969, now out of print, it states on page 27, that: 'Inserting buds on any particular side relative to compass bearings has not proved of much practical benefit in this country, but as far as possible all the buds should be inserted

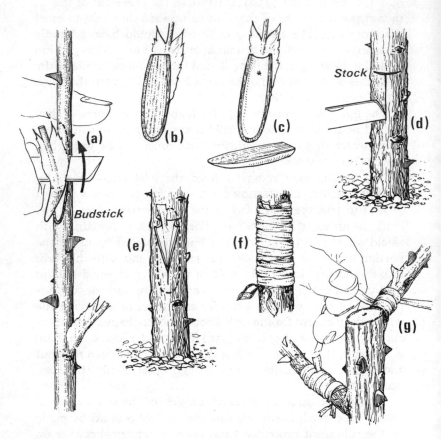

33. Rose budding – (a) removing the bud, (b) bud shield before back
wood has been removed, (c) back wood removed leaving the whole clean
except for the little knob (bud trace) in the shield, (d) making a T-cut,
(e) prepared bud inserted in the T-cut, (f) finally the bud is bound around
with raffia as shown, or adhesive tape, rubber or plastic budding tape
which is fixed with clips, (g) a standard briar rose stock after insertion
of buds (right) being tied in with raffia.

on the same side of the rootstocks in a nursery, so that they may be readily inspected and subsequently manipulated.' To the amateur who is budding several stocks that are in line this advice may be recommended.

All is now ready. Insert the bud by placing the end of the shield (beneath the eye) at the top of the T-cut prised open by the spatula-like handle of the budding knife and pushing it downwards as far as it will go (33e): any of the tail of the shield remaining above the top of the T incision may be cut off. Once the bud is in position it may be tied in securely with raffia (33f) or some other form of wrapping such as adhesive tape, rubber or plastic budding tape (which is fixed with clips), or wide polythene budding tape. If raffia is used, it should be placed in a bucket of water and wrung out before use.

Nothing more is required of the budder before the following February or early March. The stocks are then cut hard back to within 1·2–2·5 cm ($\frac{1}{2}$–1 in) of the bud. Canes or small hazel sticks should be placed to each stock as a support to which to tie the young shoot as it grows from the bud. After the stocks are cut back, growth will be very rapid, particularly in warm and moist spring weather. Keep all suckers removed.

Budding standard roses. The general procedure for budding standard roses is the same in every detail as that for bush roses except for the following points. The briar stems which form the stock should be planted the autumn previous to budding. These may be obtained from the hedgerow. Rugosa stems are also used; these can be propagated from cuttings or purchased from a nurseryman.

There is, however, one big difference in treatment of a standard rose, namely that the buds are inserted at the height at which the 'head' is to be formed – usually 1·2–1·3 m (4–4$\frac{1}{2}$ ft) for standards, 0·7–0·9 m (2$\frac{1}{2}$–3 ft) for half-standards and 1·5–1·8 m (5–6 ft) for weeping standards.

In the case of the rugosa stock, the buds are inserted into the main stem, whereas in the briar stock the buds are inserted into two or three young shoots (**33g**) allowed to grow for this purpose at the head of the stock; usually three are inserted on each standard stock, one near the base of each selected side-shoot.

Chip budding. With this method, when used on fruit and ornamental trees, there is less likelihood of losses from bud failure, which often occurs in the traditional shield – T budding. The advantage of chip budding is that the cambia (p. 118) of the scion and rootstock unite more quickly, as they are placed opposite one another, the result being that buds produce vigorous growth the following summer. Another advantage is that there is less likelihood of bud losses from canker (*Nectria galligena*). Subjects for which chip budding has been found to be successful include fruiting apple, pear and plum varieties, flowering crab apples, ornamental cherries, Norway maple (*Acer platanoides*), the red-twigged lime (*Tilia platyphyllos* 'Rubra') and *Ulmus vegeta* 'Commelin'.

Chip budding is carried out during July, August and early September. Tools needed are a sharp budding knife, sharpening equipment, a knife or secateurs for collecting the bud sticks, polythene budding tape up to 2·5 cm (1 in) wide though 1·2 cm (½ in) wide has been found most suitable, clean damp sacking in which to keep the bud sticks, labels and a waterproof pencil.

BUD WOOD. Having collected the bud wood, remove the leaves, cutting them off to leave a very short piece of leaf stalk to prevent it from pressing on the bud when it is tied. Also cut off the immature tip of the shoot. Store the prepared bud sticks in damp sacking. The budding knife, which needs to be kept very sharp, should not be used for carrying out these operations.

Rootstocks planted in autumn or winter and budded the following summer should not be cut back until the spring following budding, i.e. a two-year cycle. Select a smooth side of the stock and proceed to make a cut with the budding knife, to a depth of 3 mm (⅛ in), at an angle of about 20°, to form an

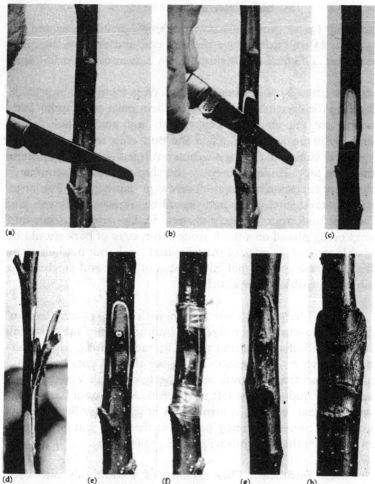

34. Chip budding – (a) preparation of stock: make a cut 3 mm ($\frac{1}{8}$ in) deep, at an angle of 20° to form a lip, (b) make a second cut of 3·7 cm ($1\frac{1}{2}$ in) above first cut, (c) remove piece of tissue to leave an inverted U, (d) removal of bud chip from scion wood by making two similar cuts, (e) fit chip so that cambia of stock and scion match as closely as possible, (f) tie the bud chip securely above and below with polythene budding tape, (g) a healed-in apple bud chip after removal of tie, (h) a union of an ornamental crab apple bud chip on an apple rootstock.

acute lip **(34a)**. Now withdraw the knife, making a second cut 3·7 cm (1½ in) **(34b)** above the first cut, and remove the piece of tissue. The final result should look like an inverted U **(34c)**.

BUD CHIP. Now remove the bud chip from the bud stick, holding the base of it towards you, then make two similar cuts, lifting out the chip between thumb and knife blade **(34d)**, then with the left hand place the bud chip in the stock – if correctly cut it will stay in position until tied with the polythene budding tape. Endeavour to fit the chip so that the cambia of the stock and scion match as closely as possible **(34e)**. For large, thick-barked stocks, the cuts should be slightly shallower, and for thin bud stocks slightly deeper. In the case of a thin bud chip being placed on a thick stock a thin strip of bark should be visible around the edge of the prepared stock, but not below the lip. This will ensure that the cambia of scion and stock are as close as possible to one another.

TYING. The bud is then tied in with a 15–23-cm (6–9-in) piece of 1·2-cm (½-in) polythene budding tape **(34f)** starting below the bud and passing round it, and finishing with a half-hitch above it. Take great care to see that all cuts below and above, apart from the bud, are properly covered. Very prominent or soft buds are best left uncovered. After about four or five weeks under normal summer conditions, slightly longer in autumn, remove the tie by pulling at the loose end of the half-hitch, when the bud chip will be healed **(34g, h)**.

Patch budding. This method of budding is one used (chiefly) for the propagation of walnuts. It is quite simple, but to be effective it is an advantage if the stock and scion (patch bud) are of an equal size and age, or near as possible. Patch budding is carried out from late May until late July, and takes with walnuts are better in hot seasons.

The actual operation calls for precision and in order to make parallel cuts, a two-bladed tool or knife is useful; such a tool

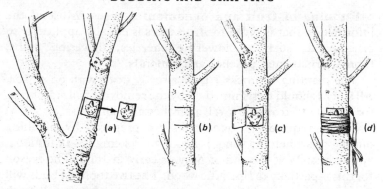

35. Patch budding – a method used (chiefly) on walnuts. To remove a patch bud (a) two parallel cuts are made horizontally, one above and one below, and two vertically either side of a scion. Three similar cuts (b) are made, while a piece of bark (c) on one side is turned back, once the patch bud is inserted cut off the piece of bark. Finally the bud is tied in (d) with raffia.

can easily be made out of a small block of wood with two saw cuts in it, into which are fixed two razor-blades which are then screwed in securely.

To prepare the bud, make two horizontal cuts with a double-razor-blade tool, and then two vertical cuts with a budding knife. The rectangular patch of bark carrying the bud is then removed (**35a**).

In the stock make two horizontal cuts and one vertical cut (**35b**). The reason only one vertical cut is made is so that the patch from the stock can be prised back as though on a hinge sufficiently to take the patch bud (**35c**), and the final vertical cut can then be made in exactly the right place, the patch filling the insertion on the stock. The bud can then be tied in securely with raffia (**35d**). It is sometimes helpful if the patch is sealed with grafting wax or a similar sealing material (p. 116).

Budding of fruit and ornamental trees. Most of the information that is given for fruit trees is equally applicable to ornamentals such as flowering cherries, flowering crabs, thorns, ornamental peaches and almonds.

The planting of stocks I have already dealt with on p. 104. All stocks should be trimmed up by the removal of all sideshoots for 0·3 m (1 ft) from soil level, after they are planted.

Shoots required as bud sticks are prepared by pinching out the tips when the shoot is about 30–38 cm (12–15 in) long, approximately at the end of May or early in June. The reason for this is partially to ripen the wood. The two topmost buds will form new shoots and in this way the sap is kept moving through the buds.

The actual operation of cutting the buds from the bud sticks is the same as for roses except for the treatment of the sliver of wood cut with the shield of bark. In apples and pears this sliver or 'back wood' is usually left intact but can be removed. Plums, cherries, peaches and nectarines, however, have the back wood removed.

The buds of plums, peaches, nectarines and cherries can be cut out a little thicker than those of apples or pears. To remove the back wood a sharp snatch from the top should be given, when it will be found that it comes away easily, leaving the little knob or bud trace attached to the shield. The shield should be at least 5 cm (2 in) long, i.e. 2·5 cm (1 in) above and 2·5 cm (1 in) below the leaf stalk. Peaches and nectarines require more careful handling than plums and cherries.

The T-shaped incision for budding fruit or ornamental trees is the same as that for roses, except for one slight differ-difference, i.e. the T is made about 7·5–10 cm (3–4 in) above ground, or several feet up when standards (stems) are budded; such trees are known as top worked.

Aftercare of budded plants. All stocks are cut down to within 20 cm (8 in) of the ground in the March following budding. As buds begin to make growth, so do the suckers, and these must be removed throughout the summer. When the bud has made 15–18 cm (6–7 in) of growth, tie up the new shoot to what remains of the stock, known as the 'snag'. Sufficient space must be left to place an ordinary lead pencil between the new shoot and the snag. If everything goes satisfactorily, the space left will be filled by the end of the summer.

Snagging. In September, the snags are removed immediately above where the bud was inserted, i.e. at the base of the new shoot. This work should not be left over until the following spring for by then the wood will have hardened and even a skilled man can have misfortune and cut through snag and shoot.

PREPARATION FOR GRAFTING

A word about tools and other equipment is essential before I begin to describe methods of grafting.

KNIVES AND STONES. The grafting knife must be strong, of good steel, and for general grafting a straight handle and blade are best, though many do use a slightly curved blade. Too much emphasis cannot be placed upon the importance of keeping the knife sharp, with a clean edge and at all times free of grit.

Knives can be sharpened on a whetstone or an oilstone. In some nurseries a sharpening stick used to be made. A piece of wood such as ash 45–60 cm (18–24 in) long and 5–7·5 cm (2·3 in) wide, was chosen, leaving at one end a portion to form a handle, sawing or cutting down the middle of the wood lengthways. On the flat surface notches were made, with a saw, in diamond fashion, about 0.6 cm ($\frac{1}{4}$ in) deep. A sharpening base was then made as follows: Russian tallow and scythe rubber dust well hammered and mixed, the proportions half and half. A little more tallow would be necessary in winter, as the cold makes the sharpening base set hard. One of the leading nurseries, at which I started my horticultural training used these sharpening sticks almost exclusively. However, today neither

Russian tallow nor sandstone scythe rubbers are easily obtainable.

TYING MATERIAL. Only the best raffia should be used, and this should be as wide and soft as possible, so that there is no fear of the raffia cutting into the stock or scion as growth of the union takes place. Today there are proprietary products on the market such as grafting tape or bandage for tying in grafts, for example Speed-Easy Fleischhauer Grafting Bandage and Rapidex British Budding Ties.

Wax. Tenax (a grafting wax) and Arbrex (a pruning and grafting compound), can be purchased from sundriesmen, or it can be made at home *if* Russian Tallow can be obtained.

396 g	(14 oz) resin
85 g	(3 oz) Russian tallow
56 g	(2 oz) Burgundy pitch
28 g	(1 oz) red ochre

This has to be made over a fire and it should be done outside, for the ingredients are highly inflammable. All that is required are six bricks and two flat iron bars to support a pail in which to make the wax. Place the resin, tallow and pitch in the pail, then light the fire, and as the mixture begins to cook, add the red ochre very carefully. The mixture must boil and be stirred occasionally until it becomes clear. When it is cooked, line a strong box with brown paper and pour the contents of the pail into it. When cool it may be broken into lumps like toffee. A wise precaution is to have a wet sack handy to throw over the pail of boiling wax if it should catch fire.

When the wax is required for grafting, lumps may be put in an old saucepan and heated up over a fire, then put on with an old paint brush when hot. Care should be exercised to keep it off the hands.

Selection of scions. All stocks required for grafting should have at least one year's growth, if a satisfactory union is to be obtained.

All wood wanted for scions should be cut off in February, the strongest year-old wood being chosen. The scions will vary considerably in size and in this way the grafter is able to choose the most suitable scion for the particular stock to be grafted. Scions are usually done up in bundles of ten. Each bundle must be labelled before being heeled or laid in the ground on the north side of a wall, fence or hedge. The bundles are laid in to a depth of 23 cm (9 in).

When they are removed for grafting, each bundle, or for that matter each scion, is washed carefully to remove any soil that may be attached and so cause the grafter's knife to lose its edge. At this stage the scions may be of any length, but when they are prepared for insertion they are usually cut down to a length of about three buds. A long shoot may, in consequence, provide material for two or more scions. Soft tips of shoots should not be used.

Preparation of stocks. Stocks for grafting may have been planted or potted; they may be failures after budding, or old fruit trees that have gone out of bearing or for which a change of variety is required. In the nursery most grafting in the case of fruit trees is on to stocks on which buds have failed. Stocks for ornamental trees and shrubs such as *Cytisus* (brooms) and rhododendrons are potted for the purpose of grafting.

There are two methods for the actual preparation of stock in readiness for grafting. One is to remove the top growth during the winter a month or more before grafting is to take place. The alternative is to remove the top growth just before grafting takes place. The latter is, I think, the better method.

METHODS OF GRAFTING

For the purpose of the amateur it will be sufficient for me to deal with the following forms of grafting, though there are many others used by experts for special purposes or for experimenting. (For those who wish to try more elaborate forms of grafting, I suggest they study *The Grafter's Handbook* by R. J. Garner, Faber and Faber.)

> Whip and tongue grafting
> Splice grafting
> Crown or rind, and cleft grafting
> Stub grafting
> Side grafting
> Bark or inverted-L grafting
> Saddle grafting
> Root grafting
> Grafting by approach or inarching
> Bridge grafting

In all grafting the vital area is that known as the 'cambium layer'. This is the thin green layer which lies between bark and wood and can be seen quite clearly on any young stem that is cut through diagonally. It is only the cells on this layer which retain the power of growing and knitting together with similar cells. In consequence in every form of grafting some part at least of the cambium of the scion must be brought into direct contact with the cambium of the stock.

Whip and tongue grafting. This is the method most used by the nurseryman for grafting his fruit and ornamental trees.

The stock is prepared by cutting it down to within 7·5–10 cm (3–4 in) of the ground level. Then a long slanting upward cut **(36a)** is made on the stock; this should be about 3·7 cm (1½ in) long. Now a small downward cut is made, which forms a tongue **(36b)** near the top of the slanting cut.

The scion is then prepared by making a similar slanting downward cut, leaving a bud midway on the opposite side of the cut. Then make a tongue **(36c)** that will correspond with the one on the stock. Fit the tongue of the scion neatly into the tongue of the stock **(36d)**; this holds the graft firmly in position. See that the cut surfaces are flush, bind the graft with moistened wide raffia **(36e)** and apply some warm grafting wax to render the

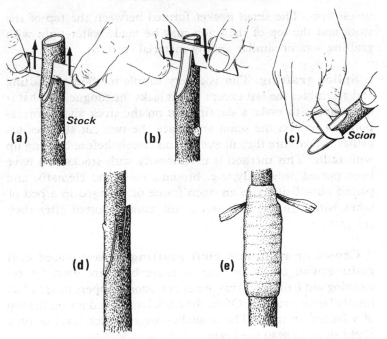

36. Whip and tongue grafting – (a) preparation of stock by making 3·7 cm (1½ in) upward slanting cut, (b) a small downward cut in stock to form a tongue near top of slanting cut, (c) first a downward slanting cut is made in the scion leaving a bud midway on opposite side of cut, then an upward cut to form a tongue to correspond with the one on the stock, (d) fit scion and stock together, (e) finally bind with raffia and seal with grafting wax or a bituminous tree dressing.

joint air- and water-tight. Or a bituminous tree dressing such as Arberex can be used. The nearer the two cut surfaces can coincide in length and width the better, but it is at least essential that the green cambia of stock and scion shall meet on one, if not both sides.

The top portion of the stock should be cut off at an angle, sloping inwards slightly. This makes a stronger joint, and provides a good support for the raffia, which is applied to make the union firm. The small pocket formed between the top of the stock and the top of the scion may be made water-tight with grafting wax or similar sealing material (p. 116).

Splice grafting. This is a very simple method of grafting and resembles the last except that it lacks the tongue. All that is necessary is to make a slanting cut on the stock and a corresponding one on the scion and place the two cut surfaces together. Make sure they fit evenly and closely before binding up with raffia. This method is used mostly with stocks that have been potted previously, e.g. brooms, roses and clematis, and placed out-of-doors in an open frame or plunged in a bed of ashes, although clematis stocks are usually potted after they are grafted.

Crown or rind, and cleft grafting. I have placed cleft grafting with crown or rind, because both are used for reworking old fruit trees that are either poor croppers or are of an unsatisfactory variety. Often the work has to be done on the top of a ladder or steps. The branches are cut back level or on a slight slope so as to shed rain.

The scions for crown or rind grafting are prepared by making either a slanting cut about 2·5 cm (1 in) long, or sometimes more, or a slanting cut with a lip which rests on the top of stock inside the rind. A slit is then cut in the rind of the same length as the scion cut. Now insert scions until either the slanting cut is flush inside the rind and lying flat against the inner wood, or the lip rests on the top of the stock. Bind with raffia in the usual way and cover with grafting wax or similar sealing material (p. 116).

Cleft grafting is a very easy method though perhaps a rather rough and ready one, yet it is quite reliable. The stock usually consists of a fair sized limb. To make the incision for the scion, a small chopper or grafting tool is required. This instrument makes a split in the limb of the stock longitudinally, and this is kept open by the pointed end or a suitable wedge. The scion is then cut, usually three or four buds in length. Two slanting cuts are made at the base of the scion in the form of a long tapering wedge. Two scions are normally inserted, their cut surfaces uniformly in contact with the cut surfaces of the wedge-like split, and the cambium layer on one side of each scion registering with the cambium layer on one side of the stock. Bind with raffia, then fill up any cavity left with clay, followed by grafting wax or similar sealing material (p. 116).

Stub grafting (37). All the lateral branches are cleared except those required to carry the scions. They should be not less than 0·6 cm (¼ in) in diameter, and can be up to 2·5 cm (1 in). The preparation of the tree can be carried out any time in the dormant season.

Each scion is cut to form a short wedge, one side of which is a little longer than the other. The lateral branch to receive the scion has a downward incision made on the upper surface about 1·2 cm (½ in) from the main branch, the cut being made nearly to the base, as illustrated. It must not, however, go farther than the centre, or the lateral branch will be weakened. To insert the scion bend the lateral down gently and push the scion in with the longer side of the wedge downwards. Then allow the lateral to spring back, so gripping the scion as in a vice. Any portion of the lateral branch that remains above the scion is cut off after insertion. All cut surfaces exposed should be sealed with grafting wax or similar sealing material (p. 116).

Side grafting (37). The scions are inserted direct into the limbs of the tree, all lateral branches being first removed – the smaller limbs are more easily grafted than larger ones. In preparing a tree for this kind of grafting, all lateral shoots may be removed, but very often it will prove wiser to leave a few lateral branches and stub graft them, in order to form a better and more balanced frame. In fact, in frameworking there is never any need to rely exclusively on one or other of the systems I

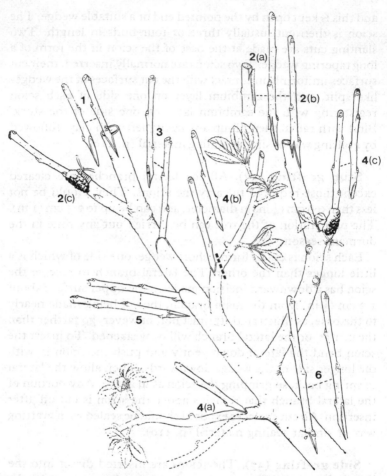

37. Frame working with its various forms of grafting. 1. prepared scion of crown or rind grafting; scion inserted, 2. whip and tongue grafting, (a) prepared stock and scion, (b) scion inserted, (c) graft tied and sealed, 3. side grafting; scion and scion inserted, 4. stub grafting, (a) a branch bent down to form a vice, (b) scion and scion inserted; once tied the rest of the branch may be cut, (c) scion and scion tied and sealed, 5. oblique side graft; scion and scion inserted, 6. bark or inverted L grafting; scion and scion inserted and fixed with a gimp pin.

describe, since all can be employed at the same time.

The scion for side grafting is cut with a longer wedge of unequal sides. The branch to be grafted is prepared by a cut being made in it at an angle of approximately 20°, and it should not be deeper than one-quarter of the branch's diameter. Again the branch must be bent back to allow insertion. After insertion, cut off the thin lip of bark behind the scion once it is secure and seal all exposed cuts with grafting wax or similar sealing material (p. 116).

In the case of the *oblique side graft* (37) the scion is prepared by making two sloping cuts approximately 2 5 cm (1 in) long, so that they form an angle on one side of about 30 to 45°. An oblique shallow cleft is then made in the branch, and this should be a little deeper than the wedge. Partly withdraw the knife and gently insert the scion, making sure that the cambium tissues of the scion and stock are in close union with each other, and seal with grafting wax or similar sealing material (p. 116).

Bark or inverted-L grafting (37). To prepare the scion, first make a long sloping cut at its base as for rind grafting. Now reverse the scion and make a similar but much shorter cut on the other side, so forming an unequal wedge. An incision is then made in the bark, in the form of an inverted L, with the angle between its two strokes opened out to about 150° instead of 90°. This L-incision should be lengthways on the branch with its shorter stroke nearest the extremity. The flap of bark formed in this way is lifted as in budding. When in position seal with grafting wax or similar sealing material (p. 116).

Saddle grafting (38). This form of grafting is one used chiefly by nurserymen for uniting named cultivars of rhododendrons to the common *Rhododendron ponticum*, which is carried out during January, February and March in heat, though it can be applied to practically any subject including fruit trees. The stocks for rhododendrons are potted twelve months before into 11·2-cm (4½–in) pots, and placed out-of-

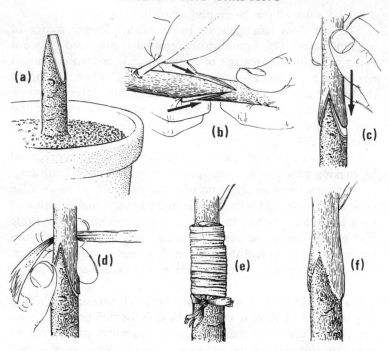

38. Saddle grafting – (a) prepared stock of *Rhododendron ponticum*, (b) making two slanting upward cuts in scion, (c) fitting scion of a named variety of rhododendron on to stock, (d) binding stock and scion with raffia, (e) finished graft, (f) callused graft.

doors in an open frame or plunged in a bed of ashes.

Scions are taken from open ground plants and it is necessary to cut into the previous year's growth. Each scion should have one good terminal bud. The scion and stock should be of even size if possible.

To prepare the scion, two slanting cuts are made at the base in an upward direction, internally, so as to form an inverted V or saddle **(38b)**. The stock is then shaped to fit the saddle by two slanting upward cuts at the top **(38a)**. When it is seen that the

stock and scion fit together **(38c, d)** they may be bound up with
raffia **(38e)**. Place in a warm closed propagating case until a
union has been made **(38f)**. It is not necessary to use grafting
wax for rhododendrons.

Root grafting. This method is used for propagating
clematis, tree peony and the double-flowered forms of *Gypsophila paniculata*.

To raise gypsophila stocks, sow seeds in March of *Gypsophila paniculata*, rather thickly so as to obtain nice long,
straight but fleshy roots. These will be used for grafting the
following February, when they may be cut into 5–7·5–cm
(2–3-in) lengths. Their thickness may vary from that of a lead
pencil to about 1·2 cm ($\frac{1}{2}$ in) diameter. The top of each stock is
cut off horizontally and the base obliquely – this is to prevent
the possibility of the scion being placed on it upside down.

Stock plants from which scions are to be taken are placed in
gentle heat to encourage young growth. When shoots are
7·5–10 cm (3–4 in) long, they are ready to be taken off as
scions.

The stocks are also prepared at the same time by the removal
of all rootlets from the portion of the root to be used. Now
split the stock down from the top for about 2·5 cm (1 in).
Carefully remove thin shavings of rind from each side of the
base of the scion. This wedge-shaped scion is then fitted closely
inside the cleft of the stock and tied up with wide raffia. Pot at
once into small pots or lay the grafts in peat moss. They are
then placed in a closed propagating case or frame or under a
propagator. Here they remain until union is complete.

An initial watering is given when first potted after grafting,
but no more until a union has been made, though a very light
overhead spraying twice daily is beneficial.

When the union and new roots have formed the young
grafted plants may be hardened off.

The amateur who has the facilities will find the grafting of
clematis intriguing. Named varieties of clematis are grafted

on to *Clematis vitalba* (Old Man's Beard or Traveller's Joy). The roots are used in the same way as gypsophila roots for stocks. Seed is sown outside in nursery beds or in pans or boxes during March.

Stock plants from which scions are to be taken are placed in a warm house in early January to encourage young growth. The shoots should be neither too firm or hard, nor too succulent. By cutting the stem between the nodes, two scions **(39b)** are made from each pair of buds where the leaves are attached.

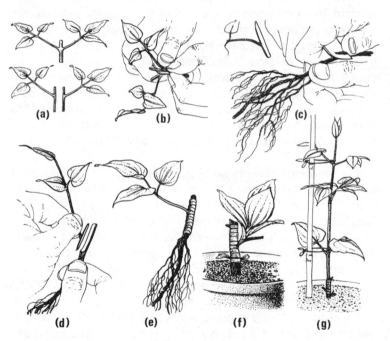

39. Root grafting – (a) stock of *Clematis vitalba* (Old Man's Beard) being prepared for grafting, (b) scions are prepared by removing a portion of a shoot above and below a node, (c) the shoot is then cut through the centre, (d) a scion is then placed on to a stock and (e) bound with raffia, (f) a young potted plant of a grafted clematis, (g) the young vigorous shoot needs to be tied as it grows.

Each scion **(39a)** should be about 2·5 cm (1 in) long with leaf attached. This leaf can be cut off, but it is not essential. The stock is prepared by making a clean cut across the top; then an upward cut **(39c)** is made the same length as the scion, leaving a slight lip at the base of the cut. Fit the scion into the incision in the stock **(39d)** and bind securely with raffia **(39e)** which should have previously been damped. No further sealing is required. Pot the grafted plant into a small pot or lay in peat, leaving the base of the bud just above soil level, and place in a closed frame or under a propagator until union has taken place. When new growth begins, the young plant may be gradually hardened off and potted into a 11·2-cm (4½-in) pot as required **(39f)**. As the young shoot grows it will need to be tied **(39g)**.

Grafting by approach, or inarching (40). By this method of grafting, the stock and scion may be united without first removing the scion from its parent. Both stock and scion

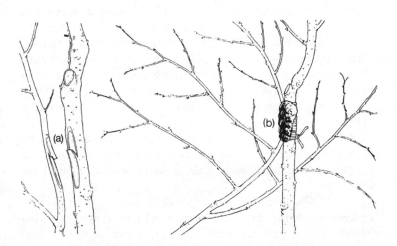

40. Grafting by approach or inarching – (a) (left) scion showing tongue (right) stock with similar tongue, (b) scion and stock bound in position and sealed with grafting wax or a bituminous tree dressing.

therefore live and grow as individual plants until a union has been made. When this has taken place, the scion is severed from its parent and henceforth lives on the stock.

Usually the stock plants are grown on in pots, though not always. I have seen young trees of *Betula pendula* (silver birch) planted as stocks around a *B. p.* 'Youngii' (weeping birch). The branches were grafted on to the stems at the required height, and when united the scions were severed from the parent tree.

A cut is made by removing a sliver of bark and wood from both scion and stock **(40a)**. A tongue should be cut in each as is done in whip and tongue grafting. The two surfaces are then spliced together, bound up with raffia and waxed **(40b)**. This method is sometimes used for camellias and magnolias.

Bridge grafting. I am not including this form of grafting as a method of increasing stock, for it is not used for that purpose but as a form of tree surgery or repair work. This method is useful if your favourite apple tree should have a portion of branch cankered or if the bark is destroyed by rabbits. To overcome this a few suitable scions of the variety may be used to form a bridge from the part above the damage to that below it. Bridge grafting can be performed by using the rind graft at both ends of the scion. It is most necessary to bind the incisions with raffia and seal with grafting wax or similar sealing material (p. 116).

In the case of a tree that has been very badly damaged by rabbits at the base of the trunk, it is sometimes necessary to place four scions on a large tree, equally spaced around the trunk.

Frame working (37). This is another method of re-working established trees and has advantages over top working, since frame worked trees crop sooner and often yield good crops in their third year after grafting.

In this method, the whole main branch system or 'framework'

of the old tree is retained. This is cleared of all the small lateral branches, except where stub grafting is used, when the scions are inserted on the laterals.

The scions are much longer than required for other forms of grafting. There should not be less than six buds to a scion, and it has been found preferable to have seven or eight buds to a scion.

Careful thought must be given to the placing of scions so that when the tree is in full growth again the newly formed lateral branches are not crowded or one-sided. Scions placed in a vertical position make stronger growth than those placed in a horizontal position. Therefore the topmost scions will often be best placed in a vertical position.

When it comes to the actual working of the tree, the six principal methods which may be applied, either separately or collectively according to convenience, are: stub, side, oblique side, bark or inverted-L, crown or rind, and whip and tongue grafting.

AFTERCARE OF GRAFTS

When the graft has made good growth, 25 cm (10 in) or more, the raffia may be loosened by gently cutting well away from the graft. This is done to prevent the raffia cutting through the rind of the scion. Do not remove raffia but leave it to fall away.

Growth will be vigorous, and sucker growths will arise from the main body of the tree as the season advances. All these should be rubbed out before they reach the length of 5 cm (2 in); anything longer must be cut out with a sharp knife. If more appear, similar treatment should be given. If the tree has been correctly grafted, sucker growths should not give trouble in later years.

✿ 6. Alphabetical list
of trees, shrubs and other plants
with methods of propagation ✿

In this chapter will be found brief but specific instructions for the propagation of a number of the trees, shrubs, herbaceous plants, alpines, annuals and biennials, etc., usually grown in gardens. In many cases the name of the genus only will be given, e.g. *Choisya* (Mexican orange blossom). In other cases more individual treatment is necessary, e.g. asparagus. Here it will be necessary to explain both the method of propagation used for *A. officinalis*, the edible species, and that for *A. plumosus* and *A. sprengeri*, both popular ornamental kinds. All the coniferous genera, such as *Abies*, *Picea*, *Taxus*, etc., are dealt with under the single heading of Conifers. Before attempting any methods of propagation I would advise that first Chapters 1 to 5 should be read.

Abelia Half-ripe cuttings in June and July, 5–7·5 cm (2–3 in) long, placed either in pots and plunged in a propagating case with a little bottom heat or inserted under a propagator, large glass jar or hand-light, or under mist.

Abeliophyllum Half-ripe cuttings taken in June and July and inserted under a propagator or under mist.

Abutilon Seeds sown in sandy soil in March in a warm house, or half-ripe cuttings in July under a propagator, hand-light or glass jar, sandy compost. All the variegated abutilons must be increased by cuttings to keep them true to type.

Acacia (Mimosa) Seeds or cuttings. Sow seeds as soon as ripe in a sandy compost under a propagator, hand-light or glass jar. Seed is the most satisfactory method, except for *A. longifolia*.

130

Acaena Seeds sown under glass in March. Division in April. Cuttings in August inserted in pots, pans or boxes in a cold frame.

Acalypha Half-ripe cuttings taken during the summer and inserted in a sandy compost in a warm propagating case.

Acantholimon (Prickly Thrift) Heel cuttings taken in July and inserted in a well drained gritty compost in a cold frame, or layering in July; or division of *A. glumaceum*.

Acanthus (Bear's Breeches) Seeds sown in spring in a cold frame. Root cuttings in late autumn or winter; or division in spring.

Acer (Maple, Sycamore) Seeds sown in pans in a cold frame immediately after gathering and removed to cool house in March. Varieties are budded or grafted on to their appropriate stocks. *A. palmatum* varieties are grafted on to *A. palmatum* in heat under glass in March. *A. negundo* 'Variegatum' is budded on to *A. negundo* outside in June.

Achillea (Milfoil, Yarrow) Seeds sown in spring in cold house or frame. Cuttings early summer or division of herbaceous varieties in autumn.

Achimenes (Hot-water Plant) The tuberous rhizomes of established plants can be separated in February and March and repotted into 12·5- or 15-cm (5- or 6-in) pots, five to six rhizomes to a pot in a compost of equal parts leafmould, loam and sand. Seeds may be sown in spring in a mixture of loam, sand and peat, in pans or pots, covering the seed 0·3 cm ($\frac{1}{8}$ in) deep and placing in a temperature of 16°C (60°F). Or soft-wood cuttings in spring in a similar temperature as for seeds.

Acidanthera The corms should be lifted in the autumn when the new corms and young cormlets may be removed and stored as for gladioli. Flowering size corms are planted in the spring. Or seeds can be sown under glass in spring.

Aconitum (Monkshood) Seeds sown in spring in cold house or frame. Division in autumn or spring.

Actaea (Baneberry) Seeds sown in a cold frame in autumn after gathering, or out-of-doors in spring. Avoid old seed.

Division of roots in spring.

Actinidia   Seeds, when available, should be sown in pots placed in a cold frame in April. Half-ripe cuttings taken in July and inserted in a propagating frame or under mist with gentle bottom heat. Alternatively hard-wood cuttings in late autumn inserted in a cold frame.

Aechmea   Offshoots or suckers, which are usually produced at the base after flowering should, when firm enough, be twisted off and potted in small pots placed in a warm moist propagating frame.

Aegle sepiaria   *see* PONCIRUS

Aeonium   Seeds sown in spring in equal parts loam, leaf-mould, and sharp sand, in well drained shallow pans, cover seed lightly, raise in a temperature of 13–18°C (55–65°F). Cuttings of shoots or leaves, which are best dried for a day in summer; insert in similar compost for seed. Or division of offsets in March.

Aesculus (Horse-chestnut)   Seeds sown immediately after gathering. If they cannot be sown at once, stratify in moist peat and sand and sow out-of-doors in spring. Rare species are best sown in pots or deep pans under glass in autumn or spring. Named varieties are whip or cleft grafted in early April. Or budded from late July to mid-August, in either case on to stocks of *A. hippocastanum*, the horse-chestnut.

Aethionema   Seeds sown in spring in pans or boxes in cold frame. Half-ripe cuttings inserted in sandy compost under a propagator, cold frame or glass jar in July.

African Corn Lily   *see* IXIA

African Hemp   *see* SPARMANNIA

African Lily   *see* AGAPANTHUS

African Marigold   *see* TAGETES

African Violet   *see* SAINTPAULIA

Agapanthus (African Lily)   Division in spring, just as growth commences. Pot in a compost of 4 parts loam, 1 part rotted manure and sand.

Agave   Seeds sown in April in a temperature of 21°C

(70°F). Offsets or suckers can be taken off at any time and potted, placing them in a propagating frame in a temperature of 16°C (60°F).

Ageratum   Seeds sown in pots, pans or boxes, in light soil under glass in a temperature of 10–16°C (50–60°F) in February and March or out-of-doors in light soil in a warm sunny spot, in April. Or half-ripe cuttings of named varieties taken in August and inserted in pots in a propagating frame.

Ailanthus (Tree of Heaven)   Take off suckers with roots attached and plant out in April. Root cuttings, inserted in deepish boxes in sandy soil at the end of February or March.

Ajuga (Bugle)   Division in March and April. Seeds can be sown in pots or boxes in ordinary seed compost in a cold frame, or soft-wood cuttings in August as for seed.

Alchemilla (Lady's Mantle)   Division in spring or autumn.

Alder   *see* ALNUS

Alexandrian Laurel   *see* DANAË RACEMOSA

Algerian Iris   *see* IRIS UNGUICULARIS

Allium (Ornamental Onion)   Most of this family increase readily by removal of bulbils and offsets from the parent bulbs in autumn or spring. Seeds may be sown out-of-doors or in a frame in spring.

Allspice, Californian   *see* CALYCANTHUS

Almond   *see* PRUNUS

Alnus (Alder)   Seeds sown in February and March in prepared bed out-of-doors. Leave seedlings in seed bed until following spring or autumn. Cuttings are moderately successful 30 cm (12 in) long inserted out-of-doors in October. Varieties are usually grafted, whip and tongue, under glass in heat in late January.

Aloe   Seeds sown in sandy soil in pots, in spring in a temperature of 16°C (60°F) or suckers in spring.

Alonsoa (Mask Flower)   Seeds sown thinly in equal parts leaf soil and loamy soil in March and April, under glass in a temperature of 13–16°C (55–60°F) or out-of-doors in late April. Soft-wood cuttings taken in March and inserted in a propagator with bottom heat in a temperature of 13–16°C (55–60°F) or half-ripe cuttings in August inserted in a propagator with gentle bottom heat.

Alstroemeria (Peruvian Lily) Seeds sown 0·6 cm ($\frac{1}{4}$ in) deep in a sandy soil, in pans in cold frame in March or May and June out-of-doors. Division of roots in April or October.

Althaea (Hollyhock) Seeds sown in June and July out-of-doors in shallow drills. They can also be treated as half-hardy annuals and sown in boxes under glass in January. Division can be used but is not advisable on account of rust. Root cuttings in winter can be used for selected varieties.

Alyssum (Madwort) Seeds sown in February and March in pans or boxes in a heated greenhouse in 10–16°C (50–60°F) or out-of-doors in April. Perennial forms, take half-ripe cuttings and insert under a hand-light or glass jar in May and June.

Amaranthus (Love-lies-bleeding) Seeds sown 0·15 cm ($\frac{1}{16}$ in) deep from January to March in a temperature of 18°C (65°F).

Amaryllis *see* HIPPEASTRUM

Amaryllis belladonna Seeds can be sown in pots or pans in a peaty soil and placed in a warm greenhouse, in spring or summer. Established bulbs are best left alone but when increase of stock is needed separate the offsets from the parent bulbs when the foliage has withered, replanting the offsets.

Amazon Lily *see* EUCHARIS

Amelanchier (Snowy Mespilus or June Berry) Seeds sown immediately they are gathered germinate readily in cool house or frame. Grafting in April on to *Sorbus aucuparia* out-of-doors.

American Cowslip *see* DODECATHEON

Ananas (Pineapple) Seeds grown in light sandy soil, in shallow pans, in a propagator with bottom heat in a temperature of 29°C (85°F); any time of the year. Also crowns, the tops of the fruit, or suckers inserted in sandy soil in a temperature of 24°C (75°F).

Anaphalis (Pearly Everlasting) Division in autumn or

134

seeds sown out-of-doors in April.

Anchusa Seeds of annuals should be sown 0·3 cm ($\frac{1}{8}$ in) deep in March in a sheltered border. *A. angustissima* is best increased from soft-wood cuttings in spring, inserted in a propagating frame, as it does not increase satisfactorily from root cuttings. *A. azurea* (syn. *A. italica*) and its varieties, root cuttings 3·7–5 cm ($1\frac{1}{2}$–2 in) long, inserted in pans or boxes 7·5–8·7 cm (3–$3\frac{1}{2}$ in) deep and placed in a cold house or frame, in autumn or early winter.

Anchusa myosotidiflora *see* BRUNNERA

Andromeda Seeds sown in late February or March on a surface of peat and sand and lightly covered with sandy peat. Place in cool house. Young tips of the branches may be used as cuttings 2·5–5 cm (1–2 in) long and inserted in a sandy peat compost under a propagator, hand-light or glass jar in August; or layers in August.

Androsace (Rock Jasmine) Seeds sown in pans or pots in light soil under glass in March. Half-ripe cuttings in sandy soil, under glass in June and July. Division in spring.

Anemone (Windflower) *A. coronaria* and *A. blanda* types, seeds sown in late spring in cool house or frame, and division after flowering or in autumn. *A. × hybrida* (syn. *A. japonica*), seeds sown in spring, division immediately after flowering or root cuttings in late autumn or winter.

Angelica (Holy Ghost) Seeds sown out-of-doors in August or as soon as possible after gathering.

Angel's Fishing-rod *see* DIERAMA

Anthemis (Camomile) Seeds sown in March and April out-of-doors or in February in gentle heat. Half-ripe cuttings in summer inserted under a propagator, glass jar or in a cold frame. Division in March.

Anthericum Liliago (St Bernard's Lily) Seeds sown as soon as ripe. Division of the roots in September.

Antholyza paniculata *see* CURTONUS

Anthurium Division of the roots in January and February placed in a propagating frame in a temperature of 21–27°C

(70–80°F). Seeds can be sown in a mixture of chopped sphagnum moss, sand and charcoal in a temperature of 21–27°C (70–80°F).

Antirrhinum (Snapdragon) *A. majus* although a perennial, is usually grown as an annual or biennial. Seeds sown in February require a temperature of 13–16°C (55–60°F) or seeds may be sown out-of-doors in April, or in July and August. All plants should have their tips pinched out when 7·5 cm (3 in) high. Soft-wood cuttings may be taken in July and inserted in a cold frame.

Aphelandra Heel cuttings of young shoots 7·5 cm (3 in) long, taken in spring, and inserted in pots of sandy soil in a warm propagator.

Apple Budding in July and August or grafting in March (see Chapter Five).

Apple of Peru *see* NICANDRA

Apricot *see* PRUNUS

Aquilegia (Columbine) Seeds sown in cool greenhouse or frame end of March, or out-of-doors April and May. Division of varieties in March.

Arabis (Rock Cress) Seeds of *A. albida* can be sown in a cold frame in March and April. Or cuttings of *A. albida* and *A. albida flore pleno* by taking young shoots in June and July, inserting them in a cold frame or a few under a large glass jar. Division after flowering.

Aralia Suckers, or root cuttings 7·5 cm (3 in) long, in March and April. These are best potted singly in 11-cm (4½-in) pots.

Araucaria excelsa (Norfolk Island Pine) Cuttings made from tips of the leading shoots, about 15 cm (6 in) long and inserted in a sandy compost in a warm propagator in September and October.

Arbutus (Strawberry Tree) Seeds of species should be sown in a cool house in a sand-peat mixture in March. Pot up seedlings singly when large enough to handle, and grow on in pots until ready for a permanent position. *A. unedo* may also be increased by layers or cuttings. Take cuttings of current year's wood from November to January and place in a propagating frame.

Ardisia Cuttings of young sideshoots taken in March and inserted in a mixture of peat and sand in equal parts, in a propagator with a temperature of 18°C (65°F). Seeds (the berries nearer the base of the plant germinate more freely) can be sown in February and March in pots or pans in a temperature of 13–16°C (55–60°F).

Aristolochia (Dutchman's Pipe)   Seeds can be sown in pots or pans in a warm greenhouse in March. Cuttings of half-ripe shoots can be rooted in a sandy soil in a warm propagator during the summer months.

Armeria (Thrift)   Seed sown in spring in a cold frame or division in spring or after flowering. Cuttings June and July under a propagator, hand-light or glass jar.

Arnebia (Prophet Flower)   Seeds sown under glass as soon as ripe or in March; or division in September; or heel cuttings in May inserted in a cold frame.

Arrowhead   *see* SAGITTARIA

Artemisia (Southernwood)   Heel cuttings in July and August in sand, using a cold or slightly heated frame.

Artichoke, Globe   *see* CYNARA SCOLYMUS

Artichoke, Jerusalem   *see* HELIANTHUS TUBEROSUS

Artillery Plant   *see* PILEA

Arum Lily   *see* ZANTEDESCHIA

Asclepias (Blood Flower)   Seeds can be sown in spring in a temperature of 10–13°C (50–55°F). Soft-wood cuttings of young sideshoots can be taken in April and rooted in a warm propagator.

Asparagus   *A. officinalis*, seeds sown in drills 3·7 cm (1½ in) deep in late March or early April in a well prepared bed or border out-of-doors. *A. plumosus* and *A. sprengeri*, seeds sown in spring in temperature of 18°C (65°F). Division in March in warm propagator.

Asperula (Woodruff)   Seeds sown as soon as ripe in late summer or autumn or in April. Or division of the roots in March and April.

Asphodel   *see* ASPHODELUS

Asphodeline   Division of the roots in spring or early autumn, being careful not to cut or damage the thong-like fleshy roots.

Asphodelus (Asphodel)   Division of the crowns in early autumn or spring.

Aspidistra   Old plants can be divided in March and April at the time of repotting.

Aster (Michaelmas Daisy)   Seeds sown in spring in gentle heat or cold frame. Division of roots in autumn or spring, using only the youngest pieces, discarding any pieces that are old or woody. Basal cuttings (often with root attached) may be potted singly in 7·5 cm (3-in) pots in autumn or spring.

Aster, China   *see* CALLISTEPHUS

Astilbe   Division in April. Alternatively roots may be divided in autumn and potted for spring planting.

Atriplex halimus (Tree Purslane)   Half-ripe cuttings inserted in an unheated frame in July or hard-wood cuttings in October and November. *A. hortensis* (Orach), seeds sown out-of-doors in March.

Aubrieta (Purple Rock Cress)   Seeds sown in spring in pans or boxes in cold frame or open ground. Division after flowering. Cuttings of young shoots in June and July inserted in a sandy compost, in an open or closed frame, or under a propagator, hand-light or glass jar.

Aucuba (Variegated Laurel)   Hard-wood cuttings 15–23 cm (6–9 in) long, taken in October and November, inserted in an open bed or frame.

Aunt Liza   *see* CURTONUS

Auricula   *see* PRIMULA

Autumn Crocus   *see* COLCHICUM

Azalea   *see* RHODODENDRON

Balloon Flower   *see* PLATYCODON

Balm   *see* MELISSA

Bamboo   Division of rootstocks in spring when there are signs of new growth. As rootstocks are very tough use a pair of strong garden forks back to back. Basal cane cuttings, 30 cm (12 in) long, of second-year canes can be taken at any time from the outer edge of a clump and inserted in a deep box of good soil and placed in a warm greenhouse. As bamboos do not flower very freely or set seed very often in the British Isles, imported seed is best, which should be sown as soon as ripe in pots filled with John Innes potting compost No 2 and placed in a warm greenhouse.

Baneberry   *see* ACTAEA

Baptisia   (False Indigo)   Division in autumn or spring. Seeds sown in pans or boxes and placed in a cold house in spring.

Barberry   *see* BERBERIS

Bartonia   *see* MENTZELIA

Bay, Sweet   *see* LAURUS

Bear's Breeches   *see* ACANTHUS

Beech   *see* FAGUS

Begonia   Seeds sown in March and April; as the seeds are minute no covering is required, just a piece of glass to cover the pot or pan, plus paper until germination. Keep in a temperature of 16–18°C (60–65°F). Seed germinates in about ten to fourteen days. Prick off in small pots when large enough to handle, using a V-notched stick as described in Chapter One, page 37. Rex and similar foliage varieties may be propagated by leaf cuttings in summer in a moist propagating case with a temperature of 16–18°C (60–65°F). Other kinds, including the winter flowering varieties, by stem cuttings in March and April using a sandy, peaty compost in a temperature of 18°C (65°F). Tuberous rooted begonias may be divided when started into growth. Fibrous rooted begonias, such as *B. semperflorens,* may be increased by seed sown in late January to early February in a temperature of not less than 18°C (65°F) or from soft-wood cuttings, when available, under similar conditions.

Bellis (Daisy)   Seeds sown in June, 0·3 cm ($\frac{1}{8}$ in) deep, out-of-doors. Plant finally in October to permanent position. Selected varieties by division after flowering.

Bells of Ireland   *see* MOLUCCELLA

Beloperone (Shrimp Plant)   Cuttings can be taken at almost any time they are available and rooted in sandy soil in a propagator with gentle bottom heat.

Berberidopsis (Coral Plant)   Half-ripe cuttings of side-shoots about 7·5–10 cm (3–4 in) long, with or without a heel, inserted in a compost of 3 parts peat and $\frac{1}{2}$ part sand in pots or pans in a warm propagator or under mist. Layering can be done in spring or autumn.

Berberis (Barberry)   This family increases very readily from seeds, which should be gathered as soon as ripe and stratified, sowing the seed the following February in boxes or pans. Cover the seed to a depth of 1·2 cm ($\frac{1}{2}$ in) and place in an open frame or direct into a prepared seed bed out-of-doors, but as seedlings vary tremendously, cuttings are preferred where trueness is required. Half-ripe cuttings should be taken in July and August and inserted in a cold frame, or glass jar or under mist. Hard-wood cuttings, 15 cm (6 in) long, with or without a heel in autumn, inserted out-of-doors or under a propagator, hand-light or glass jar. Alternatively, layers may be pegged down in April.

Berberis aquifolium   see MAHONIA

Bergamot   *see* MONARDA

Bergenia (Megasea)   Division immediately after flowering or in autumn.

Berkheya   Division in spring.

Betonica   *see* STACHYS

Bignonia capreolata   Cuttings of well matured side-shoots or root cuttings inserted in sand in a propagator with bottom heat in autumn.

Billbergia nutans   Removal of suckers, in April, by gently twisting them off the stem of the old plant; trim a few of the bottom leaves and insert in a sandy, peaty soil singly in small pots in a temperature of 21°C (70°F).

Bird of Paradise Flower   *see* STRELITZIA

Blackberry   *see* RUBUS

Black-eyed Susan   *see* THUNBERGIA ALATA

Bladder Nut   *see* STAPHYLEA

Bladder Senna   *see* COLUTEA

Bleeding Heart   *see* DICENTRA

Blood Flower   *see* ASCLEPIAS

Blood Lily   *see* HAEMANTHUS

Blood Root   *see* SANGUINARIA

Blue-eyed Grass   *see* SISYRINCHIUM

Blue Poppy   *see* Meconopsis

Bocconia   *see* MACLEAYA

Bog Myrtle   *see* MYRICA

Borage   *see* BORAGO

Borago (Borage)   Seeds, annual and perennial, sown out-of-doors in March and April. Division of perennials in April.

Bottle Brush   *see* CALLISTEMON

Bougainvillea   Cuttings of half-ripe shoots 7·5 cm (3 in) long, with a heel of older wood, in March and April, inserted in sandy soil in small pots, in a propagating case with a temperature of 18–21°C (65–70°F).

Bouvardia   Cuttings made from fairly firm young shoots in March, inserted in a sandy compost, in a propagator in a temperature of 18–21°C (65–70°F); or root cuttings in March inserted under similar conditions to top cuttings; or division in March.

Box   *see* BUXUS

Box-thorn   *see* LYCIUM

Bridal Wealth   *see* FRANCOA

Briza (Pearl and Quaking Grass)   Seeds sown out-of-doors in April where plants are to flower.

Brodiaea uniflora (syn. *Milla uniflora* and *Triteleia uniflora*)   Division of offsets in autumn when they are replanted.

Broom *see* CYTISUS

Browallia   Seeds can be sown from March to June in a greenhouse in a temperature of 16–18°C (60–65°F).

Brunfelsia   Cuttings of firm, half-ripe shoots 5–7·5 cm (2–3 in) long, inserted in equal parts peat and sand, in a propagating frame with a temperature of 16–18°C (60–65°F) or under mist.

Brunnera macrophylla (syn. *Anchusa myosotidiflora*) Root cuttings in autumn or winter in a cold frame. Seed sown 0·3 cm (⅛ in) deep in February and March under glass.

Bryophyllum   Seeds sown in sandy soil, in well drained pots or pans, cover the seed very lightly with fine soil; place in a temperature of 16–21°C (60–70°F) in March and April. Prick out plantlets which form at the margins of the leaves, or leaves can be laid flat on moist soil in a temperature of 16–18°C (60–65°F).

Buckbean *see* MENYANTHES

Buckthorn *see* RHAMNUS

Buddleia   Seeds sown from late February to early March, in pots or pans in John Innes seed compost; cover seeds lightly and place containers in a cool house or frame. *B. alternifolia* and *B. globosa*, half-ripe cuttings in June and July inserted under a propagator, hand-light or glass jar, or under mist. *B. davidii*, half-ripe cuttings in summer inserted in an unheated frame or under mist, or hard-wood cuttings in October and November in a cold frame.

Bugbane *see* CIMICIFUGA

Bugle *see* AJUGA

Bulrush Common *see* SCIRPUS

Bupleurum   Cuttings of firm sideshoots inserted in a cold frame in August, or under mist.

Burnet *see* SANGUISORBA

Burning Bush *see* DICTAMNUS or KOCHIA

Busy Lizzie *see* IMPATIENS

Butcher's Broom *see* RUSCUS

Butterfly Flower *see* SCHIZANTHUS

Butterfly Lily *see* HEDYCHIUM

Buxus (Box) Cuttings 7·5–15 cm (3–6 in) long, of mature sideshoots with heel, inserted in sandy soil in cold frame or cutting bed, in August and September. *B. sempervirens* 'Suffruticosa' (Box Edging) by division in spring.

Cactus Where heat is available seeds may be sown any time of the year, but preferably January and February. Seeds germinate very freely. When heat is used the temperature should be around 21°C (70°F). Sow in pots or pans at least 10 cm (4 in) deep. The bottom inch should be filled with crocks or burnt lumps of clay. For the next inch use similar but finer material, and let the top 5 cm (2 in) be seed compost, consisting of equal parts moss peat and silver sand passed through a fine sieve; a small portion of powdered charcoal may be added with advantage. As a rule seeds are large enough to space out 1·2 cm (½ in) apart with the point of a knife. Seeds should be covered lightly with fine sand and charcoal dust. Pots or pans must be kept nicely moist. A good plan is to stand the container in water to halfway up the sides every two or three days. If seeds are sown in heat, the water should be warm, 24°C (75°F). At lower temperatures seeds will take longer to germinate.

Propagation can also be effected by cuttings taken from May to August. Do not insert cuttings the same day as they are prepared, but the following day. Meanwhile lay them on their sides to dry. Insert in fine cutting compost either of sand and sifted peat or of sand alone, with drainage as for seeds. In the case of top-heavy cuttings use a small stick for support. The compost should be slightly damp and the pots placed in the shade for a week or ten days. They may also be increased by offshoots. In some cases grafting is practised, using such stocks as cereus and opuntia. See that the cut made in the scion and stock are of similar size, and bind up with raffia until union takes place.

Calceolaria Shrubby summer flowering species, such as *C. integrifolia* and *C. violacea*, take soft-wood cuttings in July inserted in a cold frame.

Winter flowering species such as *C.* × *burbidgei*, take soft-wood cuttings in March inserted in gentle heat.

For herbaceous summer flowering kinds, seeds are sown in June and July, in light sandy compost in pots or pans. Mix seed with silver sand, sow on top of the compost and place in cold house or frame. Cover with glass. Earlier sowings can be made from January to March under glass in a temperature of 16–18°C (60–65°F).

Calendula (Pot Marigold) Seeds sown in spring or early summer; or out-of-doors in early autumn for spring flowering. Thin out to 30 cm (1 ft) apart.

Californian Allspice *see* CALYCANTHUS

Californian Laurel *see* UMBELLULARIA

Californian Tree Poppy *see* ROMNEYA

Calla *see* ZANTEDESCHIA

Callicarpa Cuttings of half-ripened tips of shoots about 10 cm (4 in) long, trimmed beneath a leaf joint and inserted in a cold frame or under a large glass jar in June and July, or under mist.

Callistemon (Bottle Brush) Cuttings of half-ripe shoots taken in June and July inserted in a sandy peaty soil in a propagating case with a temperature of 13–18°C (55–65°F).

Callistephus (China Aster) Seeds of China asters and other bedding kinds are sown in April in gentle heat. Alternatively seeds can be sown out-of-doors in May.

Calluna (Ling) Seeds sown in February on peaty soil and lightly covered with sand. Best method of propagation is by heel or nodal cuttings, 2·5 cm (1 in) long, inserted in a sandy, peaty compost under a propagator in August out-of-doors, or in a cool house. It is not necessary to clean off bottom leaves of the cuttings. Layering can be undertaken in spring.

Caltha (Marsh Marigold) Division in spring. Plant in shady moist border.

Calycanthus (Californian Allspice) Layering in late spring, or sucker growths can be removed in early spring.

Camellia Seeds sown as soon as ripe, in boxes or singly in small pots, and placed in a warm or cool greenhouse. Use a good acid compost, i.e. a rich peaty, leafy mixture and a little silver sand. Take half-ripe cuttings of current year's wood the third week in July with or without a heel. Insert these in 2·5-cm (3-in) pots, five cuttings in a pot, under a propagator or direct in a cutting compost in a closed propagating frame with gentle bottom heat to a temperature of 13–16°C (55–60°F), or under mist. Use a compost of 3 parts silver sand, 1 part peat. Leaf cuttings with bud attached may be inserted in March, treatment being as for other cuttings. Side-grafting is, I feel, more of a job for the nurseryman than for the amateur.

Camomile *see* ANTHEMIS

Campanula (Canterbury Bell, Hare-bell) Seeds sown in June and July in boxes, open frames or borders out-of-doors. Prick out seedlings when large enough to handle 15 cm (6 in) apart into boxes or prepared beds out-of-doors. The latest sown batch may be potted up in autumn for indoor decoration in the spring.

Seeds of *C. persicifolia* and *C. pyramidalis* are sown in June and July. The former is also propagated by division in spring.

Alpine varieties by soft-wood cuttings in summer in cold frame or under a propagator.

Campsis (syn. *Tecoma*) Mature cuttings 7·5–10 cm (3–4 in) long of the current season's growth taken in July and August and inserted under mist, or in a propagating frame. Root cuttings can be cut into 5-cm (2-in) lengths and inserted singly in 6·2-cm (2½-in) pots filled with a compost of 1 part peat, 2 parts loam and 1 part sand. The pots are best plunged in ashes on a greenhouse bench where there is gentle bottom heat. Shade with newspaper, removing it when new growth commences.

Canary Creeper *see* TROPAEOLUM PEREGRINUM

Candytuft *see* IBERIS

Canna (Indian Shot) Seeds sown singly in small pots in light soil during February or March in a temperature of 18–21°C (65–70°F). Sow seeds 2·5–5 cm (1–2 in) deep. They are best soaked in water for twelve to twenty-four hours before sowing. Division of varieties just before growth commences. Start these divisions in heat.

Canterbury Bell *see* CAMPANULA

Cape Figwort *see* PHYGELIUS

Cape Fuchsia *see* PHYGELIUS

Cape Gooseberry *see* PHYSALIS

Capsicum Seeds sown in March under glass in a temperature of 16°C (60°F), using John Innes seed compost. Prick out seedlings, when large enough, singly into 5-cm (2-in) pots.

Caragana Seeds sown in pans, under glass in gentle heat in February. Layering in spring, or cuttings of half-ripe shoots 7·5–10 cm (3–4 in) long with a heel, inserted singly in small pots, or under mist in July and August.

Cardoon *see* CYNARA

Carnation *see* DIANTHUS

Carpentaria Seeds can be sown under glass, in pans, covered very lightly, in February. Or cuttings of half-ripe shoots, from growth that has been previously hard pruned, and taken with a slight heel. Insert singly in small pots in a propagator, or under mist, in July and August. Or layering in autumn.

Cartwheel Flower *see* HERACLEUM

Caryopteris Soft-wood cuttings, 5–7·5 cm (2–3 in) long, in June and July under mist, or half-ripe cuttings in August under a propagator, hand-light or glass jar.

Castor Oil Plant *see* RICINUS

Catalpa (Indian Bean Tree) Seeds can be sown under glass, no heat, in pans or boxes in February or out-of-doors in a sheltered bed in April. Or heel cuttings of half-ripe shoots inserted in a propagator with bottom heat in late July and August or under mist.

Catananche Seeds sown in April out-of-doors or in a cool house or frame in March or division in March, or root cuttings in autumn.

Catmint *see* NEPETA

Ceanothus Cuttings of spring and autumn flowering kinds. Take half-ripened wood of the current year's growth and insert under a propagator, hand-light or glass jar during July and August. Spring flowering kinds with a heel. This is most important. Or under mist.

Celosia Seeds sown in spring in well drained pots or pans filled with light sandy compost. Only just cover seed with very fine sandy soil. Place in temperature of 18–21°C (65–70°F).

Celsia Seeds of annual and biennial species sown in sandy soil in pans, in a temperature of 16°C (60°F) in March. Perennial species take cuttings in spring or autumn and root in a propagator with a little warmth.

Centaurea Biennials are sown out-of-doors in August and September and perennials in March. Annuals may be sown in spring in cool house. Division of perennials in autumn.

Centranthus *see* KENTRANTHUS

Cerastium (Snow in Summer) Seeds sown as soon as ripe or in April under glass. Cuttings in June inserted under a large glass jar or cold frame. Or division in March.

Ceratostigma or *C. plumbaginoides*, half-ripe cuttings in summer inserted under a propagator or by division in spring as growth commences. Start divisions in pots until established for planting out. *C. willmottianum*, plants are best placed in a greenhouse to encourage young growth in spring or late summer when soft-wood cuttings can be taken and inserted in a sandy soil in a propagator with slight bottom heat or under mist.

Ceropegia woodii (Hearts-entangled) Soft-wood cuttings of small sideshoots inserted in sand and peat in a warm propagator in April.

Cestrum Cuttings of half-ripe shoots 5–7·5 cm (2–3 in) long, taken with a heel any time from May to August, inserted in a sandy peaty soil in a propagator with a temperature of 16–18°C (60–65°F).

Chaenomeles (syn. *Cydonia*) (Japanese Quince) Seeds sown in April in pans, pots or boxes in light soil and placed in a cold frame, or direct in the soil of an open frame. Heel or nodal cuttings of half-ripe shoots inserted under a propagator or in a cold frame in June and July, or hard-wood cuttings in January inserted in an open frame. The earlier cuttings are to be preferred. Layers may be pegged down in June. Suckers may be removed during October to March.

Cheiranthus (Wallflower) Seeds sown out-of-doors in May and June. Alpine and semi-shrubby kinds by half-ripe cuttings in June and July inserted in a cold frame or under a propagator, hand-light or glass jar.

Chelone Seeds sown 0·15 cm ($\frac{1}{16}$ in) deep in light sandy soil in temperature of 13–18°C (55–65°F), in March or cold frame in April.

Cherry *see* PRUNUS

Cherry Laurel *see* PRUNUS

Chilean Bellflower *see* LAPAGERIA

Chilean Fire Bush *see* EMBOTHRIUM

Chilean Glory Flower *see* ECCREMOCARPUS

Chimonanthus (Winter Sweet) Cuttings of current year's half-ripened wood inserted in warm sand frame in July. Layering in spring, or seeds sown as soon as ripe in sandy compost in pots or pans placed in a cold or cool greenhouse.

China Aster *see* CALLISTEPHUS

Chinese Lantern *see* PHYSALIS

Chionanthus (Fringe Tree) Seeds sown in boxes in a cool greenhouse in February and March. Layers of vigorous shoots can be put down, giving them a slight twist, in July, August or September.

Chionodoxa Remove offsets in October.

Chlorophytum elatum and **C. comosum** (syn. *C. e. variegatum*) Very easily increased either by division in March, or the removal of the small plantlets, which grow at the end of the stems, can be potted at almost any time of the year.

Choisya (Mexican Orange Blossom) Soft-wood cuttings in July under a propagator, hand-light or glass jar, or under mist.

Chorizema Seeds sown in March, in a compost of peat, loam and sand, in well drained pots or pans in warm greenhouse. Or half-ripe heel cuttings 5–7·5 cm (2–3 in) long taken in June and July, and inserted in a sandy compost under glass.

Christmas Cactus   *see* SCHLUMBERGERA
Christmas Cherry   *see* SOLANUM CAPSICASTRUM
Christmas Rose   *see* HELLEBORUS
Chrysanthemum   *C. coccineum* (Pyrethrum) by division after flowering in September or March.

Korean chrysanthemums and early flowering outdoor varieties are also best rooted from cuttings but these are taken in March.

C. maximum (Shasta Daisy) varieties by soft cuttings in July and August or by division in autumn or spring, the latter time for *C. maximum* 'Esther Read'.

Seeds of all kinds may be sown in heat in spring but seedlings are very variable.

Indoor varieties of *C. sinense* by soft-wood cuttings 5–6·2 cm (2–2½ in) long, inserted in a sandy compost in boxes or pots, four or five in a pot, in a temperature of 7–10°C (45–50°F). Cuttings may be taken from late December until March.

Chusan Palm   *see* TRACHYCARPUS
Cigar Flower   *see* CUPHAEA
Cimicifuga (Bugbane)   Seeds can be sown as soon as ripe, in a cold frame, covering the seed very lightly. Or division of the roots in spring or autumn.

Cineraria   *see* SENECIO CRUENTUS
Cineraria maritima   *see* SENECIO CINERARIA
Cissus   Cuttings of young shoots, cut below a joint, about 5 cm (2 in) long, inserted in sandy soil in a propagator in a temperature of 21°C (70°F).

Cistus (Rock Rose)   Seeds sown in spring germinate readily but resultant plants vary tremendously. Half-ripe cuttings 5·6–7 cm (2¼–2¾ in) in July and August, inserted under a propagator or glass jar or in a cold frame, or under mist.

Citrus (Lemon, Orange, etc.)   Seeds sown 1·2 cm (½ in) deep in light sandy soil in March in a temperature of 13°C (55°F). Seedlings are of no use for fruiting but can be used as

149

stocks for budding or grafting. Budding is carried out in August; grafting in March in a temperature of 10–16°C (55–60°F).

Cladastris Seeds sown under glass, in a sandy soil, in pots, in an open propagating case with some bottom heat. Root cuttings 5–7·5 cm (2–3 in) long inserted in sandy soil in an open frame or sheltered bed out-of-doors.

Clarkia Seeds sown 0·3 cm ($\frac{1}{8}$ in) deep out-of-doors in open ground in early April. Successive sowings may be made in May and June. Where winter protection can be given, a sowing may be made in September in a cool house, in a temperature of 4–7°C (40–45°F), where they may remain throughout the winter.

Clary *see* SALVIA SCLAREA

Clematis Seeds sown in March in well-drained pans filled with a sandy compost. Place in a cold frame and cover with slate or tile to keep them dark. Germination will take at least twelve months; prick out as soon as seedlings are large enough to handle.

Half-ripe internodal cuttings of clematis species and large-flowered hybrids are best taken in mid-summer, though some propagators take them in May and as late as July. A cut is made above a pair of buds (a node) and another above the next pair of buds, and the piece of stem left is reduced to a length of about 5 cm (2 in) below the node. Now fill 8·7-cm (3$\frac{1}{2}$-in) pots with a compost of 1 part loamy soil, 2 parts peat and 3 parts sharp gritty sand. Insert four or five cuttings around the inside edge of each pot, water, and place in a cold frame – shade from bright or strong sun. The cuttings should be rooted in about six weeks, when they can be potted.

Grafting as described on p. 126 is carried out in heat in February using Old Man's Beard or Traveller's Joy (*C. vitalba*) as a stock.

Layering is another useful method of increasing large-flowered clematis. This is carried out any time during the winter up to March. Use a shoot at least eighteen to twenty-four months old – insert in 12·5–15-cm (5–6-in) pots in the ground filling with cutting compost. Now bend the stem into a U curve and insert it into the compost to a depth of two-thirds, fixing firmly with a wooden layering peg or a galvanized layering pin. In the following autumn the layers can be potted.

Cleome spinosa (Spider Flower)   Seeds sown in March in warm greenhouse; transplant seedlings when large enough to handle into 7·5-cm (3-in) pots and plant out-of-doors in May.

Clerodendrum   Root cuttings of the hardy shrub species in April, or suckers taken in spring. Also division in spring. Greenhouse species, e.g. *C. speciosissimum* (syn. *C. fallax*), *C. fragrans*; seeds may be sown in February or August in a temperature of 18–21°C (65–70°F); or cuttings of prunings 10–15 cm (4–6 in) long taken in spring and inserted in a warm propagator in a temperature of 21°C (70°F).

Clethra   Seeds sown in February to March, in boxes filled with peaty soil. The seed does not need covering. Place under glass and shade. Layering in spring or autumn. Half-ripe cuttings, 5–7·5 cm (2–3 in) long, will root readily if inserted in sand in a warm propagator or under mist in July and August.

Clianthus (Lobster Claw, Parrot's Bill)   Seeds sown in spring, or heel cuttings in spring in a temperature of 18°C (65°F), or under mist.

Clivia miniata   Seeds sown in March in a warm greenhouse. Division of suckers when repotting in February and March, using a compost of loam, leaf soil, old rotted manure and sand, and place in a temperature of 10–16°C (50–60°F).

Cobaea   Seeds sown in February and March in a warm house, or soft-wood cuttings of sideshoots in July and August in a propagator.

Cob Nuts   *see* CORYLUS

Codiaeum (Croton)   Cuttings from the ends of young shoots taken during the summer and inserted in a sandy compost in a warm propagator or under mist.

Colchicum (Autumn Crocus)   Division of corms after

the leaves have died down in June and July, when the offsets can be separated. Seeds are sown when ripe, June and July, in a prepared bed and covered with a fine gritty compost to a depth of 1·2 cm (½ in). As the leaves die down each summer, place more soil on top, and transplant the corms at the third or fourth year.

Coleus Seeds sown end of February onwards in a temperature of 18°C (65°F) – just cover seed with very fine soil. Cuttings in February where a temperature of 16°C (60°F) can be maintained throughout the winter. Cuttings can be taken at almost any time.

Colletia Well-ripened cuttings about 10 cm (4 in) long taken in November and inserted in a cold frame. Or half-ripe cuttings in summer inserted in a propagator or under mist.

Collinsia Seeds sown out-of-doors, where they are to flower, in March and April, or September. Thin seedlings when 5 cm (2 in) high to 15 cm (6 in) apart.

Columbine *see* AQUILEGIA

Colutea (Bladder Senna) Seeds sown under glass in February and March. Half-ripe heel cuttings of current year's growth taken in July and August root readily inserted in a warm propagator, or under mist.

Conifers This is a case where I make an exception to my rule of strict nomenclature, for to give individual methods of propagation for each genus would fill an entire chapter. Conifers such as *Abies, Cedrus, Chamaecyparis, Cryptomeria, Cupressus, Ginkgo, Juniperus, Larix, Picea, Pinus, Taxus* or *Thuja*, may be propagated by one or all of the following methods: (a) seeds (b) cuttings (c) grafting.

(a) Seeds must be saved from the cones. Collect these as soon as the cones commence to split open near the tips. Storage should be in a dry, airy, cool room and the temperature should not fall below about 2°C (35°F). Seeds can be sown either in autumn or spring, though spring is the better of the two. Sow end of March to about the middle of May, taking into account weather conditions and the state of the ground. The ground for outdoor sowing should be well prepared beforehand, with the addition of well rotted farmyard manure, unless the ground is very heavy, when some lightening material should be added, such as sand. Good drainage is essential. Seed beds are best just over 0·9 m (3 ft) wide, and must be broken down to a fine tilth. Sow on a calm day broadcast, or when dealing with large seeds such as those of pines, sow in rows. A covering should be given four times the thickness of the seed. Keep a sharp look out for mice and birds and set traps if necessary, or coat seeds with red lead or a combined seed dressing before sowing. The placing of netting or cotton over the bed is a wise precaution in winter against birds. Seedlings of all conifers should as a general rule remain in the seed bed for at least two years. With small quantities of seeds, sow in well drained pans, pots or boxes filled with a compost of 2 parts loam, 1 part peat and 1 part sand, and place in a cold frame.

(b) Cuttings are most satisfactory where varieties are to be kept entirely true to type. With such kinds as *Chamaecyparis lawsoniana* and its many garden forms, make the cuttings of dwarf forms from juvenile or baby growth in order to keep them dwarf. Generally speaking, cuttings are made with a heel, each cutting being 10–15 cm (4–6 in) long, or in the case of small-growing conifers such as junipers, 5–7·5 cm (2–3 in) long. They are inserted from July to September. Conifers such as *Abies* should be steeped in water at a temperature of 55–66°C (130–150°F) before being inserted in a propagator, under mist, or in an unheated frame.

(c) Grafting is carried out in heat during February and March. It is an alternative to propagation by cuttings where particular varieties must be preserved entirely true to type. As a rule seedlings of the species from which the variety is derived are used as stocks. This is a method of propagation that is not likely to be much used by the amateur.

Convallaria (Lily of the Valley) Division of the roots in September. Single crowns should be spaced 5–7·5 cm (2–3 in) apart with their points just below the surface. They enjoy plenty of leafmould and well rotted cow manure.

Convolvulus Seeds of annual species are sown in March under glass, perennial species out-of-doors in April and May. *C. cneorum*, half-ripe cuttings in June and July under a propagator.

Convolvulus major and **purpurea** *see* PHARBITIS

Coral-berried Duckweed *see* NERTERA

Coral Plant *see* BERBERIDOPSIS

Coral Tree *see* ERYTHRINA

Coreopsis Seeds of annuals and perennials sown in spring, in boxes in a cool house or frame. Soft-wood cuttings in July and August under a propagator, hand-light or glass jar. Division in spring.

Cornus (Dogwood) Seeds sown out-of-doors in March, but they may take two years to germinate. It is a good plan to soak the seeds in warm water for a few days prior to sowing. Other methods of propagation are by suckers, layers or cuttings. The suckers may be taken up in the autumn. Layering is carried out in spring, though often bushes layer themselves naturally. Hard-wood cuttings of species such as *C. alba* are made in October, 30 cm (12 in) long, and inserted in an open border, but such kinds as *C. kousa* root better from cuttings in June, 10–15 cm (4–6 in) long, made from the tips of branches and inserted under mist.

Coronilla Half-ripe cuttings 5–7·5 cm (2–3 in) long with a slight heel in June and July, inserted under a propagator, hand-light, glass jar, or under mist.

Cortaderia (syn. *Gynerium*) (Pampas Grass) Seeds easily germinated if sown in March and April. Sow seeds 0·15 cm ($\frac{1}{16}$ in) deep in a sandy soil in pots or pans that are well drained. Place under a propagator in a temperature of 13–16°C (55–60°F). Seedlings may be transplanted out-of-doors in August and September. Old clumps can be increased by division at planting time in the spring.

Corydalis Seeds of annual and perennial species can be sown in April out-of-doors where plants are to flower. Perennials can also be increased by division after flowering; and bulbous species by offsets in March.

Corylopsis   Half-ripe sideshoots taken with a slight heel, in July and August, inserted in a sandy, peaty mixture in a warm propagating frame or under mist. Layering in early spring is both easy and successful.

Corylus (Hazel)   Hazels and all other nuts such as cobs and filberts are readily increased by suckers. The rooted growths may be dug up in autumn or winter and transplanted. Seeds sown when gathered, or stratified in sand and sown in spring.

Cosmos   Seeds sown in light soil in temperature of 18–21°C (65–70°F) in March, or out-of-doors in May.

Cotinus coggygria (syn. *Rhus cotinus*) (Smoke Tree or Venetian Sumach)   Half-ripe cuttings of the current year's wood, 7·5 cm (3 in) long, inserted in a sandy cutting compost in a warm propagator or under mist in July and August; or layering in spring. This also applies to *R. cotonoides*.

Cotoneaster   Seeds should be stratified, and sown the following spring, but seedlings vary tremendously, so cuttings are to be recommended. Half-ripe cuttings 10–15 cm (4–6 in) long, inserted in sandy soil in a cold frame or under a propagator, hand-light or glass jar in June and July, or under mist. Hard-wood cuttings can also be taken in autumn and inserted in a cold frame. Also layering in spring.

Cotton Lavender   *see* SANTOLINA

Crab Apple   *see* MALUS

Crambe (Seakale)   *C. maritima*, seeds sown out-of-doors 2·5 cm (1 in) deep in March, thinned later to 30 cm (12 in) apart. Or root cuttings 10–15 cm (4–6 in) long taken in March. *C. cordifolia* (Flowering Seakale) with gypsophila-like heads of flowers. Seeds are sown out-of-doors in March. It will be three years before they flower. Or root cuttings 7·5–10 cm (3–4 in) should be taken off after flowering. Store the cuttings in sand in a frame and plant out, when leaf growth appears, in April.

Crape Myrtle   *see* LAGERSTROEMIA

Crataegus (Hawthorn, Quickthorn)   Gather berries (haws) and stratify and sow, in boxes, open frames or out-of-doors, in spring eighteen months later. Choice varieties are budded in May or grafted in April on to Quickthorn. This method is also used for *C. carrierei*, *C. prunifolia* and the double varieties.

Creeping Jenny   *see* LYSIMACHIA

Crinum   Offsets detached from the bases of old bulbs in April and May, or early autumn, and started in pots of rich loamy soil placed in a cold frame or greenhouse. Also increased by sowing the fleshy, bulb-like seeds when ripe. Place these on the surface of moist soil in a greenhouse.

Crinodendron   *see* TRICUSPIDARIA

Crocosmia (syn. *Montbretia*)   Sow seeds as soon as ripe in a pan using John Innes seed compost and place it in a cold greenhouse. Divisions of clumps after flowering; as they increase abundantly, divide them every third or fourth year.

Crocus   Increased readily by separating the young corms one from another. Seeds sown as soon as thoroughly ripe, in well drained pots or pans, and placed in a cold frame. The young seedlings should not be disturbed for a couple of years. They flower in their third season.

Croton   *see* CODIAEUM

Cucurbita   This genus includes gourds, pumpkins and vegetable marrows. Gourds, seeds sown 1·2 cm (½ in) deep in light soil in a temperature of 13–18°C (55–65°F) in April. Marrows and pumpkins, seeds may be sown in a temperature of 13°C (55°F) in light soil in April, or out-of-doors in May where they are to grow. Marrow seeds are better when kept for at least eighteen months before sowing.

Cuphaea (Cigar Flower)   Seeds sown in February and March in a warm propagating frame. Seedlings when large enough to handle can be pricked out singly into 5–7·5 cm (2–3 in) pots. Or soft-wood cuttings of young shoots taken in March and April and inserted in sandy compost in a warm propagator.

Currant, Flowering and Fruiting *see* RIBES
Curry Plant *see* HELICHRYSUM SEROTINUM
Curtonus paniculatus (syn. *Antholyza paniculata*) (Aunt Liza) Division of the clumps in October by separating the corms and replanting them 15 cm (6 in) apart. Seed can be sown as soon as ripe in pots or pans using John Innes seed compost, under glass. Alternatively sow seeds out-of-doors in the open ground in March.

Cyclamen Seeds of the greenhouse types can be sown in August and September in well drained pans or boxes, filled with light rich compost. Place seeds 1·2 cm (½ in) apart and 0·6 cm (¼ in) deep and germinate them in a temperature of 16–18°C (60–65°F). Cover each pan or box with a pane of glass followed by newspaper. Seed of the hardy cyclamen e.g. *C. coum*, *C. europaeum* and *C. neapolitanum*, can be sown, when ripe, in rich sandy soil in pots in a cold frame or unheated greenhouse.

Cydonia (Common Quince) *C. oblonga*, hard-wood cuttings in winter inserted in a prepared bed out-of-doors, or layering in spring. Varieties are grafted in March on to Quince A stocks, as used for pears.

Cynara cardunculus (Cardoon) Seeds sown in March, two seeds in a 7·5-cm (3-in) pot, in ordinary seed compost in a temperature of 13°C (55°F). Or suckers can be removed in November and stored in boxes of sand and planted out-of-doors in April. *C. scolymus* (Globe Artichoke): remove suckers in November, storing them in sand and planting them outside in April.

Cynoglossum (Hound's Tongue) Seeds sown in spring or summer in prepared beds out-of-doors. Young plants may be planted out in autumn or spring.

Cyperus alternifolius (Umbrella Plant) Seeds sown in pots or pans of light soil in a temperature of 13°C (55°F) in March and April, or division in March and April. Plants can also be increased from cuttings by shortening the leaves of the crown and leaving 2·5–5 cm (1–2 in) of the stem and inserting it in a moist, sandy compost, in a warm greenhouse.

Cyrtanthus Removal of offsets in November, or seed sown as soon as ripe in a temperature of 13 °C (55°F).

Cytisus (Broom) Seeds of species sown in February under glass germinate readily or they may be sown out-of-doors in April. Heel or nodal cuttings made in July and August and inserted in a sandy soil in a cold frame, or under a propagator, hand-light or glass jar, or under mist. Grafting of varieties on to common laburnum in heat in March (see p. 120), or layers in the open ground in May. The portion of layer to be inserted should have a cut made in it 1·2 cm ($\frac{1}{2}$ in) long and be given a half-twist before being inserted and pegged down. The half-twist is most important.

Dactylis Division of clumps in spring or autumn.

Daffodil *see* NARCISSUS

Dahlia This useful bedding plant is readily increased by seeds, though for varieties cuttings or division are essential. Grafting can be undertaken but is seldom necessary.

Seeds sown in February and March in a greenhouse with a temperature of 16–21 °C (60–70°F) will germinate readily.

To obtain cuttings, place the tubers in boxes with just sufficient soil to cover them to their necks. Start them into growth in a greenhouse with a temperature of 7–10 °C (45–50°F) in February. Give a good soaking and wait for young shoots to appear. When 7·5 cm (3 in) high they may be taken off either with a heel or without. Only use the best, discarding all weakly ones. Insert singly in small thumb pots in a sandy compost. Keep in same temperature until rooted.

Division may take place as soon as growth commences. See that each portion of tuber has at least one shoot. The divisions may be potted or boxed in John Innes potting compost No 1, or a mixture of peat, sand and soil in equal parts, to grow on.

Grafting is carried out in a similar heat to that recommended for cuttings. It is done in February and a wedge graft is used. A cutting is shaped into a wedge at the base, inserted into a portion of tuber and bound with raffia. I do not think the amateur will need this method unless it is for the novelty and experience.

Daisy *see* BELLIS

Danaë racemosa (Alexandrian Laurel)   Division in March and April or seeds sown in pans with ordinary seed compost, in a cold frame, in March.

Daphne   Seeds sown as soon as ripe in pans or boxes and stood in a cold frame. Grafting in early spring under glass, using as stocks either *D. laureola* or *D. pontica*. Layering is ideal for *D. cneorum* and *D. genkwa*; the latter can also be raised from root cuttings in winter, or under mist. *D. burkwoodii* and *D.* 'Somerset', half-ripe cuttings 5–7·5 cm (2–3 in) long in June to early August, inserted in a propagator, or under mist. *D. mezereum*, sow seed as soon as ripe, as above; or nearly ripe cuttings 15–23 cm (6–9 in) long taken the first week of September. Use hormone powder. Insert in a compost of equal parts loam, coarse sand and leafmould, and place in a cold frame. Root cuttings in autumn and winter. *D. odora* and *D. o.* 'Aureomarginata', half-ripe cuttings of current year's growth about 7·3 cm (3 in) long, in late July. Trim each cutting at a node or leaf joint; insert in a propagator in a greenhouse, or under mist. Or graft in February on to stocks of *D. laureola*, in a propagator in a warm greenhouse.

Datura   Shrubby species by cuttings 15 cm (6 in) long inserted in sandy soil in a propagating case with bottom heat, in spring or autumn, in a temperature of 21 °C (70°F). Annual species, seeds sown in light sandy soil, covered 0·3 cm ($\frac{1}{8}$ in) deep, in a temperature of 13 °C (55°F), March and April.

Davidia (Dove Tree or Pocket-handkerchief Tree)   Seeds, as they have a very hard skin, must be stratified and sown the following March, under glass. Cuttings of half-ripe shoots, 7·5–10 cm (3–4 in) long and taken with a heel, should be inserted in a propagating frame with bottom heat in July and August, or under mist. Layering can be undertaken in early spring.

Day Lily *see* HEMEROCALLIS

Deadnettle *see* LAMIUM

Decaisnea   Seeds sown under glass in ordinary seed

compost in February. Or freshly gathered seeds can be sown in autumn in a cold frame.

Delphinium   Seeds sown in spring under glass, or out-of-doors in prepared beds or frames in early summer. Cuttings and division in March. Cuttings inserted in a sandy soil in a cold frame root readily, but must be made from young shoots before they become hollow at the base. Shade for first few days and afterwards during bright sunlight.

Dendromecon   Half-ripe cuttings of short shoots inserted in sandy soil, singly in pots, and placed in a propagating frame, in July and August. Root cuttings in December or spring under glass, in an open propagator.

Desfontainea   Seeds sown under glass in February and March, in a compost of 2 parts peat and 1 part sand. Cover seed lightly. Half-ripe cuttings 7.5 cm (3 in) long, with a heel, inserted under a hand-light or large glass jar.

Deutzia   Seeds of species may be sown in March in a cold frame, or half-ripe cuttings inserted in sandy soil under a propagator, hand-light or glass jar in June and July, or under mist. Hard-wood cuttings can be taken in October and inserted in a prepared bed out-of-doors. Cuttings should be used for selected varieties as seedlings vary somewhat in character.

Dianthus   This genus includes perpetual and border carnations, picotees, and pinks. Seeds of all dianthus are best sown in the spring, though they may be sown at nearly any time of the year where facilities permit. Perpetual-flowering carnation seedlings are best potted into 5-cm (2-in) pots instead of being pricked off into pans or boxes first. All selected varieties of perpetual-flowering carnations should be increased by cuttings. Prepare these in December, January and February from side-shoots formed midway up the flowering stems and insert in sand or sandy soil under glass.

Border carnations and pinks may be increased in June and July by cuttings prepared from non-flowering basal shoots, in sandy soil in a frame or out-of-doors, but as a rule border carnations are increased by layering in July as explained in

Chapter Three, pp. 58–60. *D. barbatus* (Sweet William), sow seeds out-of-doors in May and June.

Dicentra (syn. Dielytra) (Bleeding Heart) Division in February, March and April, or root cuttings in March and April placed in a cold frame. Cuttings are inserted in sandy soil in a temperature of 13°C (55°F).

Dictamnus alba (Burning Bush) Seeds should be sown as soon as ripe, out-of-doors in the open ground. Division in spring or root cuttings in November and December.

Dielytra *see* DICENTRA

Dierama (Angel's Fishing-rod, Wand Flower) The offsets of the bulbous root stock can be separated in October when they can be planted in a nursery bed until they reach flowering size in the following autumn.

Diervilla *see* WEIGELA

Digitalis (Foxglove) Seeds sown in spring or summer out-of-doors. Spring sowing is best.

Dimorphotheca Annuals – sow seeds in March and April under glass or mid-April out-of-doors where the plants will flower. *D. barberae* and *D. ecklonis* sub-shrubby species, take half-ripe cuttings in September, inserted in a cool house or cold frame where frost is excluded in winter.

Dipelta Half-ripe cuttings, taken from a stock plant growing under glass. Insert cuttings in a peat and sand compost, in a propagator in May and June, or under mist.

Dipladenia Cuttings 2·5–5 cm (1–2 in) long inserted in moist peat and sand compost in a warm propagator in March and April, in a temperature of 16–18°C (60–65°F).

Dodecatheon (American Cowslip) Seeds sown under glass in March or out-of-doors in April. Division of roots in March or September.

Dog's-tooth Violet *see* ERYTHRONIUM

Dogwood *see* CORNUS

Doronicum (Leopard's Bane) Division of roots in September or March.

Dove Tree *see* DAVIDIA

Draba   Seeds sown as soon as ripe under glass in March and April or division of roots in April.

Drimys   Cuttings of half-ripe shoots 10 cm (4 in) long inserted in a warm propagating frame in July and August, or under mist. Layering in March and April.

Drumstick Primula   *see* P. DENTICULATA

Dryas (Mountain Avens)   Seeds can be sown as soon as ripe in pans or pots in a cold frame. Division of the roots in March.

Dutchman's Pipe   *see* ARISTOLOCHIA

Easter Cactus   *see* SCHLUMBERGERA

Eccremocarpus (Chilean Glory Flower)   Seeds sown 0·15 cm ($\frac{1}{16}$ in) deep in pots or pans filled with a light sandy soil, well drained, under glass in a temperature of 13°C (55°F) in March and April. Or cuttings of ripened shoots inserted in sandy compost in a warm propagator in autumn.

Echinacea   Seeds sown in spring in light soil in temperature of 10–13°C (50–55°F). Root cuttings in October, placed in a cold frame, or division in autumn or spring.

Echinops (Globe Thistle)   Seeds sown in spring. Root cuttings in late autumn or winter placed in a cold frame, or division in autumn or spring.

Edelweiss   *see* LEONTOPODIUM

Edraianthus   Seeds sown in a gritty seed compost in March, or soft-wood cuttings inserted in pans of sandy compost from May to July, placed in a cold frame.

Elaeagnus   Layering out-of-doors in September and October. Half-ripe cuttings inserted in a propagator with gentle heat in July or under mist, or hard-wood cuttings in November inserted in a cold frame.

Elder   *see* SAMBUCUS

Elm   *see* ULMUS

Embothrium (Chilean Fire Bush)   Seeds sown, in February, in pans under glass where there is moderate bottom heat. Root cuttings about 3·7 cm (1$\frac{3}{4}$ in) long, can be taken in December. Insert the cuttings horizontally and singly in small pots, filled with a compost of 2 parts peat, 1 part loam and 1 part sand. Place pots in an open propagating frame with moderate bottom heat, or under mist.

English Iris *see* IRIS XIPHIOIDES

Enkianthus Seeds sown in a peaty compost, lightly covered, in February. Soft-wood cuttings about 5 cm (2 in) long can be taken June and July and inserted in a warm propagating case or under mist. Layering in March and April, using the tips of shoots of the previous year's growth. Give each shoot a slight twist and peg down in the normal way. Rooting takes about eighteen months before they are ready for lifting.

Epimedium Division of roots in August and September.

Eremurus (Foxtail Lily) Seeds should be sown as soon as ripe in a made-up frame under glass, without heat. Small quantities are best sown in pots or pans. They are slow to germinate and seedlings will often take up to six years before they flower. Plants can be divided, but as the fleshy, octopus-like roots are easily broken, propagation by seed is preferable. In any case, plants are best left undisturbed for at least three years.

Erica (Heather) Seeds sown in February on peaty soil and lightly covered with sand. Best method of propagation is by heel or nodal cuttings, about 2·5 cm (1 in) long, inserted in a sandy, peaty compost under a propagator in August out-of-doors, or in a cool house, or under mist. It is not necessary to clean off the bottom leaves of the cuttings. Layering or division can be undertaken in spring.

The Cape Heaths, cuttings of greenhouse species, about 2·5 cm (1 in) long, require a temperature of 16°C (60°F) and are best taken in spring. The double pot method is excellent.

Erigeron Seeds sown in spring out-of-doors. Division in autumn or spring.

Erinacea (Hedgehog Broom) Seeds sown in ordinary seed compost in February and March. To speed up germination seed is best refrigerated or frozen before being taken into sudden warmth. Half-ripe cuttings in July and August, inserted in a compost of 3 parts sand and 1 part peat moss in a sun frame or under mist. Some propagators prefer to take more fully ripened cuttings in October.

Erinus (Summer Starwort)   Seeds sown in March under glass, or division in autumn.

Eriobotrya (Loquat)   Soft-wood cuttings taken in spring. To encourage suitable growth for cuttings, older plants should be cut back in the previous autumn and taken into a warm greenhouse in February. Insert cuttings in a sandy-peaty compost or a soilless compost singly in small pots or under mist.

Erodium (Heron's-bill)   Seeds sown in February and March, in ordinary seed compost, in pots or pans under glass in a temperature of 7°C (45°F). Or heel cuttings of half-ripe shoots taken in June and July inserted in a sandy compost in pots in a cold frame. Or division of roots in April.

Eryngium (Seaholly)   Seeds sown in April and May in a cold frame. Root cuttings in autumn or winter placed in a cold frame, or division in autumn or spring.

Erythrina (Coral Tree)   Soft-wood cuttings in the spring when young shoots are about 7.5 cm (3 in) long, taken with a heel of the old wood. Insert in sandy soil in a propagating frame or under mist.

Erythronium (Dog's-tooth Violet)   Sow seeds in a cold frame in August. Thin out seedlings when large enough to handle, then leave them in the frame until September three years later. Or increase them by offsets taken off in August.

Escallonia   Seeds sown in March, covered lightly, and placed in a cool house; or half-ripe cuttings in July and August inserted in a sandy compost under a propagator, hand-light or glass jar, or under mist.

Eschscholtzia   Seeds sown 0.15 cm ($\frac{1}{16}$ in) deep in August and September, or April and May, out-of-doors in a sunny border.

Eucalyptus   Seeds sown in early spring in pans of light soil in a temperature of 18°C (65°F).

Eucharis (Amazon Lily)   Division of offsets in spring;

after they are divided, water very sparingly. Seeds (when obtainable), if well ripened, can be sown in early spring, in a high temperature; seedlings however will take much longer than the offsets before they flower.

Eucomis (Pineapple Flower) Division of offsets in March. If seeds are well ripened they can be sown in spring in a rich gritty soil. It will take four to five years before bulbs reach flowering size.

Eucryphia Half-ripe cuttings of sideshoots 5·6–7·5 cm (2¼–3 in) long, taken with a heel, inserted in a sandy peaty soil in a propagator or under mist, in late June or early July. Layering can be undertaken in late July and August – all that is needed is a twist of the stem. Layers should be ready for lifting in the autumn twelve months afterwards. Seeds can be sown under glass in February, but hybrids will not come true to type.

Euonymus (Spindle Tree) Seeds sown in March in pans or boxes and placed in a cool greenhouse or cold frame. Layering in early summer. Hard-wood cuttings, taken with a heel, 7·5–10 cm (3–4 in) long, inserted in a cold frame in October and November. *E. japonicus*, half-ripe cuttings in July and August or early September. *E. fortunei radicans* can be increased by division, or cuttings as for *E. japonicus*.

Euphorbia (Spurge) The hardy spurges can be divided in spring or autumn, though *E. wulfenii* is increased either from seed (I have found this spurge self-sows itself freely in my garden) or soft-wood cuttings in spring. *E. fulgens*, cuttings inserted in a warm propagator in spring and summer. Soft-wood cuttings of *E. pulcherrima* (Poinsettia) are taken with a heel in April. Insert singly in small pots in half sand and half peat in a temperature of 21°C (70°F).

Evening Primrose *see* OENOTHERA
Everlasting Flower *see* HELICHRYSUM, HELIPTERUM
Everlasting Pea *see* LATHYRUS LATIFOLIUS
Exacum Seeds can be sown in March though they are best sown in August, for flowering the following year, in a sandy, peaty soil in a temperature of 16°C (60°F). Soft-wood cuttings

of young shoots can be taken in March and inserted in sandy soil in a propagator with bottom heat, or under mist.

Exochorda (Pearl Bush) Seeds should be gathered when ripe and stored in dry sand until they are sown in boxes of ordinary seed compost, under glass in February. Soft-wood cuttings taken 5–7.5 cm (2–3 in) long with a slight heel are inserted in a warm propagator, or under mist in July. It is also sometimes possible to remove sucker growths with a piece of root attached, in autumn.

Fagus (Beech) The common beech produces a good crop of seed after a favourable spring and when this happens seedlings can easily be dug up in beech woods. Otherwise seeds can be sown out-of-doors in early spring and covered to a depth of 1.2 cm ($\frac{1}{2}$ in). Leave the seedlings in the seed bed for eighteen months from the time of sowing.

Named varieties are grafted on to the common beech, *F. sylvatica* in spring, either under glass or out-of-doors.

False Acacia *see* ROBINIA

False Hellebore *see* VERATRUM

False Indigo *see* BAPTISIA

False Solomon's Seal *see* SMILACINA

Fatshedera Cuttings of sideshoots or tips 10–12.5 cm (4–5 in) long, inserted in equal parts of peat and sand in pots or unheated propagating frame, in July and August. Pot rooted cuttings in 10–11 cm (4–4$\frac{1}{2}$-in) pots the following spring.

Fatsia Seeds are sown singly in pots in the spring and germinated in a cool greenhouse. Half-ripe cuttings 10–15 cm (4–6 in) long can be taken in August. Insert singly in small pots. Place in a propagator with bottom heat.

Ferns Division, spores, or plantlets.

Division of ferns, whether hardy or greenhouse, is very simple, the operation being carried out in March and April. Large clumps of hardy ferns may be divided easily by the method of placing two large garden or hand forks back to back through the clumps as explained on p. 44. Greenhouse ferns are divided in a similar manner when growth commences in February and

41. The fine brown dust-like bodies beneath fern leaves are known as spores and under the right conditions sporelings can be raised.

March, and potted in a compost consisting of 3 parts turfy loam, 1 part flaky leafmould, and 1 part coarse sand. The texture of the compost should be open and in a rough state, not finely broken up, as this will tend to clog.

Spores are usually plentiful. They are to be found on the undersides of the leaves (41), and in appearance they are like fine brown dust. The most satisfactory method of raising sporelings is to choose lumps of fibrous loam and peat, sterilize them with scalding water, place them in glass jars with lids that fit closely and, while the lumps are still saturated, scatter the spores over the surface and close the lids tightly. The jars are then stood in a shady corner of the greenhouse and left undisturbed. No further watering should be required for the less evaporation there is the better. First signs of germination will look something like liverwort. The tiny green mass of sporelings should be separated into small patches and transferred into well drained pans filled with sifted loam and peat and kept in a close frame until growth becomes active. Eventually the young plants are pricked out singly. Tepid rain water should be used at all times when moisture is required.

Asplenium bulbiferum and *Polystichum setiferum* 'Proliferum' can be propagated by plantlets which form along the mature fronds. Take such a frond and peg it down in a pan or box filled with light soil, until roots are formed or, alternatively, take off the young plantlets that appear on the top surface of the leaf and place round the sides of a 12·5-cm (5-in) pot filled with light soil. In both cases pot off singly when large enough to handle.

Ficus elastica (Rubber Plant) Small pieces of stem with leaf attached can be inserted singly in small pots filled with sandy soil, in a propagating frame with bottom heat in March, or under mist. Plants that have grown too large can have their tops taken out and rooted like the stem cuttings, in a temperature of 10°C (50°F). Or air layering can be used **(18)**. To do this, cut a 1·2-cm ($\frac{1}{2}$-in) wide ring of bark off the stem, dust the bare piece of stem with hormone rooting powder, bind with sphagnum moss and cover with a piece of polythene, securing it firmly top and bottom with adhesive tape. When roots show through remove below the ball of moss and very carefully pot the new rooted plant. *F. carica* (Common Fig), layers can be bent over and pegged down into the ground during the summer; remove rooted layers in autumn and grow on until ready for planting in a permanent position.

Fig, Common *see* FICUS CARICA

Figwort *see* SCROPHULARIA

Filbert Nuts *see* CORYLUS

Filipendula Division in autumn or seeds sown in pans or boxes, under glass in autumn.

Firethorn *see* PYRACANTHA

Flax *see* LINUM

Flowering Bamboo of Japan *see* NANDINA

Flowering Currant *see* RIBES

Flowering Nutmeg *see* LEYCESTERIA

Flowering Seakale *see* CRAMBE

Forget-me-not *see* MYOSOTIS

Forsythia Soft-wood cuttings 10–15 cm (4–6 in) long inserted in a warm sand frame in June, or hard-wood cuttings 30 cm (12 in) long inserted out-of-doors in October. The latter method is preferred.

Fothergilla Half-ripe cuttings 7·5–10 cm (3–4 in) long taken with a heel and inserted in a sandy compost in a propagator with bottom heat or under mist, in July, root readily, though the easiest method is layering in August. Choose shoots of the current year's wood and after forming a tongue, peg down the layers into the ground to which peat and sand have been added.

Four o'clock Plant *see* MIRABILIS

Foxglove *see* DIGITALIS

Foxtail Lily *see* EREMURUS

Francoa (Bridal Wreath) Seeds sown in March and April in pots of fine sandy, peaty soil in a temperature of 13°C (55°F). Soft-wood cuttings can be taken in June and inserted in a close propagating frame under glass. Or division in March.

Freesia This popular greenhouse plant is easily propagated by offsets, which are freely produced. These are planted in early August with about 2·5 cm (1 in) of soil above the top of the corm, and will flower from early December onwards. A 23-cm (9-in) pot will hold 10 corms, using J.I. No 2 potting compost. Supports will be required. Seeds may be sown in early April in 15-cm (6-in) pots filled with light well-drained soil, or the J.I. compost as given for corms. Space the seeds 2·5 cm (1 in) apart and 1·2 cm ($\frac{1}{2}$ in) deep, and place in a temperature of 16°C (60°F). Usually seedlings appear in about four to five weeks. These should be grown on in a cool temperature and as much light and air as possible. In this way the seedlings may be left without transplanting and some seedlings may bloom during late summer, but in general flowering will start in

October. Seeds sown in late summer will bloom next spring.

Fremontia *see* FREMONTODENDRON

Fremontodendron (syn. Fremontia) Seeds sown in March in small pans in a temperature of 7–10°C (45–50°F). Pot off into small pots as soon as large enough to handle.

French Honeysuckle *see* HEDYSARUM

French Marigold *see* TAGETES

Fringe Tree *see* CHIONANTHUS

Fritillaria Increased easily by offsets in the autumn, or by seeds sown when ripe in a gritty compost of loam, leaf soil and sand in a cold frame. Seedlings take four to six years before they flower.

Fuchsia This is an easy plant to propagate. Young soft-wood cuttings taken in April may be used for greenhouse kinds. Cuttings should be about 7·5–10 cm (3–4 in) long and trimmed below a node. Insert in boxes or 7·5-cm (3-in) pots filled with sharp sand, or half sand and peat, which can be stood in a larger box with glass covering the whole, or in a frame, in a temperature of 16°C (60°F).

Cuttings of hardy kinds such as *F*. 'Riccartonii' can be taken in June and inserted in a warm propagator, or taken as late as August and inserted in a cold frame, glass jar or propagator.

Seeds of all kinds can be sown in spring in pans of light soil and germinated in a temperature of 16°C (60°F). Prick off into small pots when large enough to handle.

Funkia *see* HOSTA

Gaillardia Seeds can be sown in heat in February or in June out-of-doors in a warm sheltered border, the seedlings being protected throughout the winter and planted out in spring. Root cuttings taken in the autumn readily form new plants, or soft-wood cuttings may be inserted in sandy soil in a cold frame in July.

Galantus (Snowdrop) Division by offsets after flowering which is the best time for replanting.

Galega (Goat's-rue) Division in autumn or spring. Half-ripe cuttings taken in the summer and inserted in a cold frame or a few under a large jam jar. Although seed germinates freely sown in a cold frame in October, vegetative methods are best as seedlings are often of poor quality.

Galtonia (Summer Hyacinth) Seeds can be sown as soon as ripe in the open, or under glass in rich gritty soil in spring. It will be two to three years before bulbs attain flowering size. Offsets can be taken off older bulbs in spring or autumn and replanted.

Gardenia Half-ripe cuttings 5–7·5 cm (2–3 in) long, taken with a heel in February or March. Insert cuttings around the edge of a 7·5-cm (3-in) pot, well-drained, filled with a sandy, peaty mixture. Place in a propagating frame with a temperature of 21 °C (70 °F). As rooting starts, give slightly cooler conditions. When rooted, pot singly in 7·5 cm (3-in) pots in John Innes compost No. 1. For potting on as the plants grow, use equal parts loam, peat, sand, well rotted manure and a little charcoal.

Garrya Sow seeds in pans or boxes in March in a cool greenhouse to produce both male and female plants. For male plants, which produce the finest catkins, grafting is necessary, which is done during February on *Aucuba japonica*, under glass in heat. Layering in June is also possible.

Gaultheria Division in spring, planting in a mixture of peat and sand. Seeds sown in February in boxes of finely sifted peat, under glass or out-of-doors in April. Layering in September and October: all that is needed is a sharp twist before inserting layers in the soil to which should be added peat and sand.

Gazania Cuttings of sideshoots inserted under a propagator or in a cold frame from July to late August. They require protection from frost during the winter. Seeds sown in March in heat in well drained pots or pans filled with light sandy soil.

Genista Seeds sown in March under glass germinate readily, or they may be sown out-of-doors in April. Heel cuttings made in August, inserted in a cold frame, or under a propagator or glass jar in sandy soil, or under mist.

Gentiana Seeds should be sown as soon as ripe in pans

filled with a porous soil and placed in a frame. Germination will be hastened by the snow treatment as described in Chapter One, Section (II) p. 26. As gentians resent too much root disturbance, they are best pricked out singly from the seed pan direct into small pots. Many species and varieties can be divided in spring or early autumn, but only do so when necessary. In the case of *G. sino-ornata*, young rooted pieces can usually be detached without seriously disturbing the parent plant.

Geranium Seeds of hardy species and varieties may be sown in early March in pans or boxes filled with sandy soil and placed in a cold frame or greenhouse. Alternatively they may be sown out-of-doors late March and April in a sunny border. Division of roots in autumn or spring.

Geranium–Ivy-leaved, Regal, Scented-leaved and Zonal *see* PELARGONIUM

Gerbera This beautiful plant is rather exacting. Seeds are best sown as soon as ripe in boxes filled with John Innes compost, very lightly covered and placed in a temperature of 10°C (50°F). Germination will take from seven to fourteen days. As soon as seedlings are large enough to handle, they should be pricked out. When approximately 7·5 cm (3 in) high, pot off into 7·5-cm (3-in) pots in John Innes potting compost. Plants from spring-sown seed may be planted out in May, but plants from summer- or autumn-sown seed must have protection from frosts. Old seed seldom germinates. Established plants may be divided in spring. The ideal spot for gerberas is a raised south border.

Germander *see* TEUCRIUM

Gesneria Leaf cuttings can be taken with a 1·2 cm ($\frac{1}{2}$-in) piece of stalk, inserted in equal parts peat and sand, in a temperature of 21°C (70°F) in a propagating frame in May and June.

Geum Seeds sown in February and March in boxes in a cool greenhouse or frame, or late spring to early summer out-of-doors in prepared beds. Division in autumn or spring.

Gladiolus Varieties are propagated by separation of corms. The old corm withers, leaving the new corm, or corms, on top, with cormlets attached. These cormlets may be planted to form flowering-size corms in two to three years. Cormlets should be lifted when foliage turns brown. Store in dry, frost-proof building and plant out in mid-April.

Seeds can be sown in April, 2·5 cm (1 in) apart, 1·2 cm (½ in) deep, in boxes no less than 12·5 cm (5 in) deep, or in prepared beds in the garden, or in frames in rows 15 cm (6 in) apart.

Globe Amaranth   *see* GOMPHRENA

Globe Artichoke   *see* CYNARA SCOLYMUS

Globe Flower   *see* TROLLIUS

Globe Thistle   *see* ECHINOPS

Globularia cordifolia   Seeds sown under glass in June when ripe or in February. Cuttings in August and September, inserted in sandy soil beneath a large glass jar or in a cold frame. Division in September—either pot or plant out at once.

Gloriosa   Seeds sown singly in small pots, using a sandy compost, in January, under glass where there is bottom heat and a moist atmosphere. Offsets can be removed when growth starts in spring. Careful handling is necessary.

Gloxinia   *see* SINNINGIA

Goat's-rue   *see* GALEGA

Golden Rod   *see* SOLIDAGO

Gomphrena (Globe Amaranth)   Seeds sown in February and March, covered to a depth of 0·3 cm (⅛ in) under glass, in a temperature of 18°C (65°F).

Gooseberry   *see* RIBES

Gorse   *see* ULEX

Gourd   *see* CUCURBITA

Grape   *see* VITIS

Grape Hyacinth   *see* MUSCARI

Grevillea (Silk Oak)   *G. robusta,* seeds sown in March and April in a sandy soil and germinated in a temperature of 16°C (60°F). As the seeds are large and flat they should be placed either on their points or sideways, but not flat. Cuttings root easily in spring in a temperature of 10–16°C (50–60°F), or under mist. *G. rosmarinifolia* and *G. sulphurea,* half-ripe cuttings taken in

July and inserted in a sandy/peaty compost in a close frame with bottom heat, or under mist.

Griselinia   Hard-wood cuttings, in October and November, made from sideshoots with a small heel and inserted in a compost of 3 parts loamy soil, 1 part peat and 1 part sand, in a cold frame; or half-ripe cuttings in July under mist.

Gunnera   Seeds sown in boxes of peaty soil and placed on the ground beneath the large gunnera leaves; leave the boxes there until the seedlings are large enough to pot or plant out into a prepared bed. Or division in spring.

Gynerium   *see* CORTADERIA

Gypsophila   Seeds of perennials may be sown in gentle heat under glass in March, or out-of-doors in April and May.

Annuals are sown in April and May out-of-doors. Soft-wood cuttings of *G. paniculata* 'Bristol Fairy' in July, inserted in a cold frame or under a propagator or glass jar, in sandy soil root readily. *G. paniculata* 'Bristol Fairy' can also be propagated from young shoots if they are grafted on to stocks of *G. paniculata* in February and March, as described on p. 125.

The alpine gypsophilas, e.g. *G. fratensis* and *G. repens* and varieties, may be divided in early summer, or soft-wood cuttings may be inserted under a propagator, hand-light or glass jar in June and July.

Haemanthus (Blood Lily)   Division of offsets when re-potting in early spring.

Halesia (Snowdrop Tree)   Seeds need to be stratified before sowing out-of-doors in March. Cover seeds to a depth of 1·2 cm ($\frac{1}{2}$ in). Or layer in early spring at leaf bud burst. Select shoots of the previous year's growth; the bend in the layer should be about 23 cm (9 in) from the end of the shoot. Half-ripe cuttings 5–7·5 cm (2–3 in) long taken with a heel, in May and June and inserted in a sandy/peaty compost beneath a large glass jar or under mist.

Hamamelis (Witch Hazel)   Seeds sown in spring in a cool house. Germination will take up to two years. A better method is to graft on to seedling stocks of *H. virginiana* in March in heat.

Or layering in early spring: a tongue should be made 20–25 cm (8–10 in) from tip of shoot of previous year's growth; insert layers at least 10 cm (4 in) below soil and tie portion of layer above soil to small cane.

Hare-bell *see* CAMPANULA

Hare's-tail Grass *see* LAGURUS

Hawthorn *see* CRATAEGUS

Hazel *see* CORYLUS

Hearts-entangled *see* CEROPEGIA

Heather *see* ERICA

Hebe (Shrubby Veronica) Half-ripe cuttings 7·5–10 cm (3–4 in) long inserted in a cold frame or under a propagator or large glass jar in August, or under mist.

Hedera (Ivy) Half-ripe cuttings 10–15 cm (4–6 in) long, inserted in a cold frame during August and September, or hard-wood cuttings inserted in an open frame in November.

Hedgehog Broom *see* ERINACEA

Hedychium (Butterfly Lily) Division of the rhizomatous roots in March at potting time; use a rich soil and place in a moderate temperature.

Hedysarum (French Honeysuckle) Seed sown in spring under glass, or half-ripe cuttings in summer inserted beneath a large glass jar or under mist. Or shoots of the previous year's growth can be layered in early spring. A twist should be made about 15 cm (6 in) from the end of a shoot.

Helenium Seeds sown in spring in cool house. Soft-wood cuttings taken from the base of the plant in June and July, and inserted in a cold frame or under a propagator, or division in autumn or spring. Choice new varieties may be divided in autumn, potted into 7·5-cm (3-in) pots and overwintered in a cold frame.

Helianthemum (Sun Rose) Seeds sown in pans or boxes in sandy soil in cold house or frame in March. Half-ripe cuttings with or without a heel in July and August, inserted in a cold frame, or under a propagator or large glass jar.

Helianthus Seeds sown in pans, pots or boxes in April.

H. annuus (Sunflower), seeds are sown 5 cm (2 in) deep out-of-doors in April. Division in autumn or spring of perennial kinds. *H. tuberosus* (Jerusalem Artichoke), by separation of tubers, January to March.

Helichrysum (Everlasting Flower, Immortelle) Seeds of the hardy and half-hardy annuals sown thinly in light soil in pans or shallow boxes, covering the seed to a depth of 0·3 cm (⅛ in), in February and March in a temperature of 16–18°C (60–65°F), or in a cold frame in April. *H. serotinum* (syn. *H. angustifolium*) (Curry Plant), seeds sown in pots or pans under glass in February. Cuttings of half-ripe shoots, in June and July, about 5 cm (2 in) long, inserted in sandy cutting compost in a cold frame or under a large glass jar. *H. bellidioides* by division in spring.

Heliotrope   *see* HELIOTROPIUM

Heliotropium (Heliotrope)   Seeds sown in February and March in gentle heat, or soft-wood cuttings in spring or early autumn in pots, pans or boxes, in a light sandy soil, placed in a close propagating frame. Spring cuttings are best.

Helipterum (Everlasting, Immortelle)   Seeds sown under glass, in light soil 6 mm (¼ in) deep, in a temperature of 13–16°C (55–60°F) in March.

Helleborus (Christmas Rose and Lenten Rose)   Seeds sown 0·3 mm (⅛ in) deep in shallow pans or boxes filled with sandy soil, in cold frame, in July or March. Division of roots in March.

Hemerocallis (Day Lily)   Division of roots in early autumn or spring.

Hepatica   Division just as the leafless plants are coming into flower. *H. angulosa* flowers in February and March and *H. triloba* and its varieties in March and April. Never divide in summer or autumn.

Heracleum   *H. mantegazzianum* (Cartwheel Flower), seeds sown in a cool greenhouse or frame in March or by division in March or October.

Heron's-bill   *see* ERODIUM

Hesperis (Sweet Rocket) *H. matronalis*, although a perennial, is best treated as a biennial and seeds may be sown out-of-doors in April o·6 cm (¼ in) deep, in a sunny position. Double flowered kinds may be divided carefully in spring, or cuttings may be taken from July to September and inserted in a sandy soil under a hand-light or propagator or glass jar in a shady spot.

Heuchera Division of roots in spring. Or seeds can be sown in March and April in a cold frame.

Heucherella Division of roots in spring.

Hibbertia Half-ripe cuttings about 7·5 cm (3 in) long, inserted in a propagating case under glass in May, or under mist.

Hibiscus *H. rosea sinensis*, hard-wood cuttings taken in autumn, inserted in a propagating frame with bottom heat. *H. syriacus* and its varieties, take half-ripe cuttings of short sideshoots with a slight heel in July and insert under a propagator or glass jar or hand-light, or under mist. They may also be grafted on to *H. syriacus* in heat in February.

Seeds of greenhouse varieties may be sown in February, in a sandy peaty mixture in well drained pots, in a close frame or under a propagator in a temperature of 24°C (75°F).

Himalayan Honeysuckle *see* LEYCESTERIA

Hippeastrum (Amaryllis) Seeds sown in March 1·2 cm (½ in) apart in well drained pots filled with rich, gritty, sandy soil and covered with o·6 cm (¼ in) of soil. The pots or pans should then be plunged in peat moss litter in a temperature of 16–18°C (60–65°F), each being covered with a pane of glass. Germination takes from ten to fourteen days. Prick off when large enough, either singly in small pots or twelve in a 15-cm (6-in) pot. It will be about three years before they flower.

Hippophaë (Sea Buckthorn) As one of the chief beauties of *H. rhamnoides* is its berries, it is necessary to propagate female forms by layering in autumn. Seeds should be stored in dry sand through the winter, and later they can be sown out-of-doors in spring, or in boxes in an open frame.

Hoheria Half-ripe cuttings 7·6–10 cm (3–4 in) long

inserted in a sandy, peaty soil under a large glass jar or cold frame, or under mist, in June and July. Or layers of the previous year's growth put down in April. A twist should be made about 23 cm (9 in) from the tip of each shoot. Rooted layers can be lifted twelve months later.

Holly *see* ILEX

Hollyhock *see* ALTHAEA

Holodiscus Seeds sown in February in pans or pots, in ordinary seed compost; cover the seeds lightly and place in a cold frame. Cuttings of half-ripe shoots in June and July inserted in a cold frame, or hard-wood cuttings inserted out-of-doors in the autumn.

Holy Ghost *see* ANGELICA

Honesty *see* LUNARIA

Honeysuckle *see* LONICERA

Hop *see* HUMULUS

Horse-chestnut *see* AESCULUS

Hosta (syn. *Funkia*) (Plantain Lily) Division of the roots when new growth starts, which will be in March and April.

Hot-water Plant *see* ACHIMENES

Hound's Tongue *see* CYNOGLOSSUM

Houseleek *see* SEMPERVIVUM

House Lime *see* SPARMANNIA

Hoya (Wax Plant) Cuttings of the previous year's growth taken with a piece of stem about 10 cm (4 in) long, with a pair of leaves attached; insert in a sandy soil in a propagating frame with bottom heat of 18°C (65°F). Or layers can be taken; lower a shoot and remove some leaves, inserting this portion of the shoot in a pot filled with a sandy-peaty compost, which should be kept moist until the layers are rooted.

Humea (Incense Plant) Seeds can be sown in June, July and August, in John Innes seed compost, put through a fine sieve; cover seed very lightly and place in a warm greenhouse with a temperature of 13°C (55°F). Pot on seedlings singly in small pots as soon as they are large enough to handle. During the winter, keep plants almost dry.

Humulus (Hop)   Seeds sown 0·3 cm ($\frac{1}{8}$ in) deep in pots under glass in April, or out-of-doors in May.

Hyacinth   *see* HYACINTHUS

Hyacinthus (Hyacinth)   Seeds are sown when thoroughly ripened in September, in boxes or pans filled with light sandy soil, covered to a depth of 1·2 cm ($\frac{1}{2}$ in). Seedlings can be left in a cold frame during the winter but must be protected from frost. It will take from four to seven years before the bulbs will flower.

Offsets can be removed from the old bulb when it is lifted, planting them 5 cm (2 in) deep in sandy soil. To encourage bulbils, slit the base of the bulb at the end of the first year's growth. Plant the bulbils in a rich sandy soil. They will flower in three to four years.

Hydrangea   This popular greenhouse and hardy shrub is easily propagated. Nodal cuttings 7·5 cm (3 in) long should be taken during May, June and July. Choose fairly stocky cuttings from the base of the plant. To provide these a few stock plants may be taken into the greenhouse in early summer. Weak or thin cuttings are useless. For the amateur one cutting may be inserted in each 7·5-cm (3-in) pot, but for larger quantities a box may be used. Place pots or boxes in a close propagating case, or the cuttings can be placed directly in a propagating case or frame. When rooting takes place air may be given, and the young cuttings potted into 7·5-cm (3-in) pots, using a good loamy mixture of 1 part soil, 1 part peat and 1 part sand. I have also rooted prunings pushed in the soil as markers for seed rows. *H. paniculata*, half-ripe cuttings, under glass in July and August inserted under a hand-light or large glass jar. All hydrangea cuttings can be rooted under mist.

Hymenanthera   Hard-wood cuttings 3·7–5 cm ($1\frac{1}{2}$–2 in) long, taken with a heel, inserted in a normal cutting compost in a cold frame, October and November. Or half-ripe cuttings, with a heel, inserted in a sandy-peaty compost in a propagating frame in July, or under mist.

Hypericum (St John's Wort)   Sow seeds thinly in March

in cool house or frame and cover lightly. Soft-wood cuttings 5 cm (2 in) long inserted in a cold frame, glass jar or warm propagator in July and August. Hard-wood cuttings inserted either in a cold frame or out-of-doors. *H. calycinum* may be divided in spring.

Hyssop *see* HYSSOPUS

Hyssopus (Hyssop) Seeds sown out-of-doors in April or half-ripe cuttings 5–7.5 cm (2–3 in) long, taken with a heel and inserted in a cold frame in June and July; or division of the plants in spring or autumn.

Iberis (Candytuft) Seeds sown either in boxes in March, or out-of-doors in open ground in April. Half-ripe cuttings in July and August under a propagator or glass jar, or in a cold frame.

Idesia Seeds can be sown under glass in the spring in John Innes Seed Compost covering the seed by 0.6 cm ($\frac{1}{4}$ in). Or by layering in spring using shoots of the previous year's growth; give a twist 15 cm (6 in) from the tip of the shoot and peg the layer into the ground, to which sand and peat have been added. Or half-ripe cuttings taken with a heel and inserted in a propagating frame in May and June, or under mist.

Ilex (Holly) Seeds when gathered should be stratified as recommended in Chapter One, p. 19. After eighteen months they may be sown in March in pans or boxes, or direct into a frame, filled with a sandy seed compost. Cuttings taken in September can be successfully rooted under mist. Hard-wood cuttings inserted in an open frame in autumn will, however, take longer to root.

Illicium Layers can be put down just before new growth starts in early spring; choose shoots of the previous year's growth, and after making a cut beneath the shoot 15 cm (6 in) from its tip, peg the layer into the soil to which peat and sand have been added. Choose a sheltered border. Pot the rooted layers eighteen months later.

Immortelle *see* HELICHRYSUM, HELIPTERUM

Impatiens (Busy Lizzie) Greenhouse species by seeds

sown 0·3 cm ($\frac{1}{8}$ in) deep in a light soil in March and April in a temperature of 13–18°C (55–65°F). With the hardy species seeds can be sown out-of-doors in April where plants are to flower. Cuttings of all kinds root readily at any time between March and August if inserted in a light sandy compost in a warm propagating case. Cuttings of *I. holstii* and *I. sultanii* root readily in water.

Incarvillea Seeds sown in pans or pots in March in a temperature of 13°C (55°F), or in a cold frame in April. Division in autumn or spring.

Incense Plant *see* HUMEA

Indian Azalea *see* RHODODENDRON INDICUM and R. SIMSII

Indian Bean Tree *see* CATALPA

Indian Shot *see* CANNA

Indigofera Seeds sown in John Innes seed compost in pots or pans and placed in a cold frame in February and March. Half-ripe cuttings taken with a heel 5–7·5 cm (2–3 in) long, inserted singly in small pots in a sandy-peaty compost and placed in a propagating frame or under mist in June and July. Or root cuttings, about 2·5 cm (1 in) long; place the cuttings horizontally in small pots filled with a sandy-peaty compost, covering them with 0·6 cm ($\frac{1}{4}$ in) of a fine sandy compost. Place in an open propagating frame in a greenhouse in December.

Inula Seed can be sown under glass in February, in pots, pans or boxes or out-of-doors in a sheltered bed in March and April. Or division of the roots in October or March.

Ipomoea *see* PHARBITIS

Iresine Soft-wood cuttings can be rooted at almost any time in a warm greenhouse, though slightly firmer cuttings taken in August and inserted inside the edge of a 12·5-cm (5-in) pot root readily in a warm propagating frame.

Iris This is a very large and varied genus and one which includes some species with fibrous roots, and some with bulbs. Propagation of the various species and forms varies accordingly.
 The iris popularly known as the flag iris may be increased by

division of the rhizomes immediately after flowering, usually July and August. Growth buds attached to the rhizome may be detached when the size of a pea and inserted as cuttings. They should be cut out of the rhizome with a sharp penknife. Where new varieties are wanted, seeds may be sown as soon as the seed-pods turn brown, in pans, pots or seed boxes 10 cm (4 in) deep at least. Good drainage is essential. Use a compost of 1 part fibrous loam, 1 part flaky leafmould passed through a 1·2-cm ($\frac{1}{2}$-in) sieve, 1 part granulated peat; do not sow deeper than 1·2 cm ($\frac{1}{2}$ in). Plunge containers in sand or ashes in a cold frame during the winter.

Fibrous rooted irises such as *I. sibirica* and *I. kaempferi*, may be divided in spring.

Bulbous irises such as *I. reticulata*, *I. xiphium* (Spanish Iris), and *I. xiphioides* (English Iris), are readily increased by separating the bulb clusters, in late spring or early summer in the case of *I. reticulata*, and late summer and early autumn for the Spanish and English. *I. unguicularis* (syn. *I. stylosa*) (Algerian Iris), division in September (not spring as often advised); it is essential that the Algerian Iris has an ample supply of moisture after planting.

Isatis tinctoria (Woad) Seeds sown in spring in a prepared bed out-of-doors; it is biennial. *I. glauca* (perennial), division in March or October at planting time.

Isoloma Seeds sown in March, in a sandy-peaty soil in a temperature of 24°C (75°C), or cuttings inserted in a similar soil mixture and temperature in a propagating frame.

Itea Seeds, when available, can be sown under glass in April. Half-ripe cuttings, in August and September, made about 7·5−10 cm (3−4 in) long, with a heel, and inserted in cutting compost in a warm propagating frame or under mist. Suckers can be taken off in autumn or early spring.

Ivy *see* HEDERA

Ivy-leaved Geranium *see* PELARGONIUM

Ixia (African Corn Lily) Seeds sown in autumn in pans and placed in a cold frame; leave the seedlings to grow on in the pan for the first year, afterwards plant out in a well drained soil in a south-facing border; lift annually when foliage has died down and replant in October in cold districts, and November in warmer climates. Seedlings will take three to four years to flower. Offsets can be removed in October and replanted and the corms will flower two years afterwards.

Ixora Half-ripe, short-jointed cuttings taken in spring or summer, inserted singly in small pots filled with a sandy-peaty compost and plunged in a propagating frame with a temperature of 21°C (70°F).

Jacaranda Seeds sown in pots filled with a light sandy compost, in February in a temperature of 24°C (75°F). Cuttings of half-ripe shoots, taken during the summer, inserted in a sand and peat compost in a propagating frame with a temperature of 24°C (75°F); shade until rooted. Or under mist.

Jacobinia Half-ripe cuttings 5–7·5 cm (2–3 in) long, taken with a slight heel in May and June, inserted in pots filled with a sandy-peaty compost and placed in a propagating frame with a temperature of 24°C (75°F). Or under mist.

Jacob's Ladder *see* POLEMONIUM

Jamesia Cuttings of well ripened shoots 5–7·5 cm (2–3 in) long, taken in the autumn and inserted in sandy-peaty soil in a cold frame, or beneath a large glass jar. Or half-ripe cuttings in May, inserted in a propagator with bottom heat or under mist.

Japanese Quince *see* CHAENOMELES

Jasione Seeds sown out-of-doors in a prepared bed in April, or division of roots in October and November.

Jasmine *see* JASMINUM

Jasminum (Jasmine) Hard-wood cuttings of *J. nudiflorum* 10–18 cm (4–7 in) long with a heel, in November, inserted in a cold frame.

J. officinale half-ripe cuttings 7·5–12·5 cm (3–5 in) long with a heel if possible, in July and August, inserted in a close frame.

Jeffersonia dubai (syn. *Plagiorhegma dubai*) Seeds can be sown as soon as ripe in July and September, in sandy soil in pots or pans; cover the seed 1·2 cm ($\frac{1}{2}$ in) very lightly and place in a cold frame. Division of plants in August or March.

Jerusalem Artichoke *see* HELIANTHUS TUBEROSUS

Jerusalem Sage *see* PHLOMIS

Jew's Mallow *see* KERRIA

June Berry *see* AMELANCHIER

Kaffir Lily *see* SCHIZOSTYLIS

Kalanchoe Seeds sown in March in an open gritty seed compost; as seed is very fine no covering is required. Germinate in a temperature of 16°C (60°F). Cuttings also root easily in a warm greenhouse. After removal leave the cuttings to dry for a day or so before inserting them in a sharp sandy soil; water very moderately until rooted.

Kalmia Seeds sown in February and March in pans or boxes filled with a finely sifted sandy-peaty compost – seeds should not be covered – and placed in a greenhouse where there is gentle bottom heat. Cuttings of half-ripe shoots of current year's growth, 5–7·5 cm (2–3 in) long, inserted in sandy-peaty soil in a cold frame or under a large glass jar in August. Or under mist. Pot rooted cuttings the following spring.

Kansas Feather *see* LIATRIS

Kentranthus ruber (syn. *Centranthus ruber*) (Red Valerian) Seeds sown out-of-doors in March and April; or division in November or March.

Kerria (Jew's Mallow) *K. japonica* 'Pleniflora', and *K. japonica*, hard-wood cuttings 15–23 cm (6–9 in) long, in October and November out-of-doors. Or division of sucker growths in early spring.

Kirengeshoma palmata When good seed can be obtained, sow as soon as ripe, in pots or pans filled with a mixture of equal parts decayed leafmould and peat moss, with a little coarse sand or John Innes compost No. 1 plus extra leafmould. Place containers in cool or cold greenhouse. Prick out seedlings when large enough to handle into 7·5- or 8·7-cm (3- or 3$\frac{1}{2}$-in) pots. Or division in early spring before new growth starts. Old plants, however, do not divide very easily.

Kniphofia (Red-hot Poker) Seeds sown in spring in pans or boxes in a cool house or frame. Division of roots in spring just as they show new signs of life. Do not divide in autumn.

Kochia (Burning Bush, Summer Cypress) Seeds sown in pots or boxes in a warm greenhouse during March, or out-of-doors in early summer. Prick off into boxes or separate pots, and plant out in a sunny aspect during May or early June.

Koelreuteria Seeds sown under glass in boxes filled with John Innes seed compost, covered to a depth of 1·9 cm ($\frac{3}{4}$ in), February and March. Root cuttings 3·7 cm ($1\frac{1}{2}$ in) long, inserted horizontally and singly in small pots filled with a compost of 2 parts loam, 1 part peat and 1 part sand. Place pots on bench, plunged in peat or ashes in a warm greenhouse. Cover pots with paper, which should be removed when the cuttings show signs of growth.

Kolkwitzia Half-ripe cuttings inserted in a propagating case, with some heat in July and August, or under mist.

Korean Chrysanthemum *see* CHRYSANTHEMUM

× **Laburnocytisus adamii** Graft on to seedling stocks of common laburnum (*L. anagyroides*) in spring.

Laburnum Seeds of true species may be sown thinly in March in boxes or frames.

Varieties such as *L.* 'Vossii', *L.* 'Aureum', etc., are grafted on to common laburnum (*L. anagyroides*) in spring. Young roots of these stocks are also used for grafting the varieties of cytisus. *L.* 'Vossii', also increased from hard-wood cuttings of current season's growth, 23–30 cm (9–12 in) long, inserted out-of-doors in late November and December.

Lachenalia Seeds sown as soon as fully ripe in pots or pans of very gritty soil in a temperature of 16°C (60°F). Prick out seedlings when large enough to handle, grow on in sunny place near to the glass. Seedling bulbs will flower in about two to three years from time of germination. Offsets can be potted on in September to flower the following spring.

Lady's Mantle   *see* ALCHEMILLA

Lagerstroemia (Crape Myrtle)   Cuttings of young shoots inserted in a sandy compost root readily if placed in a warm propagating frame in March and April.

Lagurus ovatus (Hare's-tail Grass)   Seeds should be sown out-of-doors in April.

Lamb's Ear   *see* STACHYS

Lamium (Deadnettle)   The ornamental deadnettle roots almost wherever it touches the ground; it can be divided at any time of the year.

Lampranthus (syn. *Mesembryanthemum*)   Seeds sown 0·3 cm ($\frac{1}{8}$ in) deep in pans filled with sandy soil in March in a temperature of 13–18°C (55-65°F). Prick out seedlings when large enough to handle or annuals may be transplanted in June out-of-doors. Cuttings or firm young shoots may be inserted in sandy soil in a frame or cool greenhouse from June to August.

Lantana   Seeds sown under glass in February and March in a temperature of 16°C (60°F) germinate readily. Soft-wood cuttings taken from old stock plants which have been cut back root easily if taken in late spring or early summer and inserted in a propagating frame, where there is gentle bottom heat.

Lapageria (Chilean Bellflower)   Layering in spring or early summer is the most successful method of propagating lapageria. Layers should be made from the previous year's growth and either a small tongue made on the underside of the shoot or a sharp twist made about 15 cm (6 in) from the tip of the shoot; insert layers at this point, in a mixture of peat and sharp sand in a seed box. To keep the soil moist cover compost with damp sphagnum moss.

Lathyrus (Sweet Pea)   Seeds may be sown in the first or second week in October in boxes or pots, usually 11-cm (4½-in) size, placing two to four seeds in each pot, or one or two in a 7·5-cm (3-in) pot, removing the weakest. Stand in a cold frame all winter. Alternatively seed may be sown in boxes under glass in January. For outdoor sowings where plants are to flower, the second and third weeks of March are best. Cover seed to a depth of 2·5 cm (1 in). Seeds may also be sown out-of-doors in the autumn in some districts, where cloches are available for protection. *L. latifolius* (Everlasting Pea), sow in spring or summer in pots filled with sandy seed compost and place in cold frame. Or cuttings of young shoots taken in April and inserted in a cold frame or division in March.

Laurel, Alexandrian   *see* DANAË RACEMOSA

Laurel, Californian   *see* UMBELLULARIA

Laurel, Cherry   *see* PRUNUS LAUROCERASUS

Laurel, Portugal   *see* PRUNUS LUSITANICA

Laurel, Variegated   *see* AUCUBA

Laurus nobilis (Sweet Bay)   Hard-wood cuttings inserted in sandy soil in a cold frame, or under a propagator or hand-light, during August, September and October, or under mist. Shade until rooted.

Lavandula (Lavender)   Cuttings, either heel or nodal, of ripened wood 5–7·5 cm (2–3 in) long, inserted in a sandy soil in a cold frame in August. Heel cuttings can also be torn off old plants in the spring and inserted in a cold frame or bed in the open ground without preparation to the cutting itself; an Irishman's or Dutchman's cutting!

Lavatera   Seeds of biennial species are sown in April, May and June out-of-doors.

Seeds of *L. olbia* are sown in spring and placed in a cool house; or half-ripe cuttings may be inserted in sandy soil under a hand-light, propagator or glass jar in July. Or under mist.

Lavender   *see* LAVANDULA

Lavender, Cotton   *see* SANTOLINA

Layia elegans (Tidy Tips)   Seeds sown out-of-doors where they are to flower, in March and April.

Lemon   *see* CITRUS

Lemon-scented Verbena   *see* LIPPIA

Lenten Rose   *see* HELLEBORUS

Leonotis (Lion's Tail)   Soft-wood cuttings with a slight

187

heel in June and July, inserted in a compost of 2 parts loamy soil, 1 part peat and 1 part sand, in a propagating case with bottom heat.

Leontopodium (Edelweiss) Seeds sown in spring in a compost of loamy soil, leafmould and sharp sand, on granite chippings in pans or pots which should be well crocked; place them out-of-doors in a cool shady position.

Leopard's Bane *see* DORONICUM

Leptospermum (Manuka or Tea Tree) Half-ripe cuttings with heel, sideshoots 5–7·5 cm (2–3 in) long, inserted in a sandy-peaty compost in a propagator with some bottom heat, or beneath a large glass jar; or under mist. Hard-wood cuttings of the current year's growth with heel attached can be taken in October and November and inserted under a large glass jar in a prepared bed out-of-doors.

Lespedeza Soft-wood cuttings taken in the spring or half-ripe cuttings in early August; in either case insert in normal cutting compost in a propagating case where there is some bottom heat.

Leucojum (Snowflake) Removal of offsets as soon as the foliage withers and the bulbs ripen. Usually bulbs need to be divided every four to five years.

Leucothoë Seeds can be sown under glass, in February and March, in finely sifted peat; cover the seeds lightly. Half-ripe cuttings taken with a heel, 5–7·5 cm (2–3 in) long, inserted in a sandy-peaty compost, and placed in a cold frame or beneath a large glass jar, or under mist.

Lewisia Seeds sown in March and April in pans, which must be well drained and filled with a compost of sandy loam, peat and finely broken brick, and placed in a cool house. Division of the roots in March or April.

Leycesteria (Flowering Nutmeg, Himalayan Honeysuckle) Seeds sown in March in pans or boxes in a cold frame germinate readily. Hard-wood cuttings may be taken in October. Bushes once established will seed themselves in and around the parent plant.

Liatris (Kansas Feather) Seeds sown as soon as ripe out-of-doors 0·15 cm ($\frac{1}{16}$ in) deep, in a light sandy soil, in August and September. Seedlings will take three to four years to bloom. Basal buds can also be removed in the spring and potted on, or the old plants can be divided and replanted in March and April.

Libertia Seeds sown as soon as ripe in pans of light soil, August and September. Division of the rhizomes in spring.

Ligustrum (Privet) Hard-wood cuttings 30 cm (12 in) long, in October inserted in prepared bed out-of-doors.

Lilac *see* SYRINGA

Lilium (Lily) Seeds sown in autumn when quite fresh give good germination. Seeds may also be sown in spring, but the former season is the better time because some varieties if sown in spring will probably lie dormant for twelve months. No growth from lily seeds (that have germinated) will be shown the first year. Soil should consist of 2 parts well rotted fibrous loam, 1 part oak or beech leafmould (well rotted), 1 part good sharp sand and 5 parts charcoal. Cover the seeds with 1·2 cm ($\frac{1}{2}$ in) of sifted compost mixed with a little finely broken charcoal. Use pots, pans or boxes and cover with a slate or asbestos sheet until seeds germinate. Seeds may also be sown in prepared beds out-of-doors. Scales may be taken off mature bulbs after flowering or in the autumn and placed in boxes of similar compost to that recommended for seeds. In time they will form new bulbs (see p. 90). Aerial bulblets or bulbils in the axils of the leaves may be taken off after flowering and placed in boxes, and treated like seeds, being covered to a depth of 5 cm (2 in). By deep planting and subsequent earthing up of the stems to 7·5–10 cm (3–4 in) just prior to flowering, it will be found that a large number of bulblets will be formed along underground stems and these may be treated as for scales. Division of normal bulb clusters in autumn provides a simple, but usually rather slow, method of increase.

Lily *see* LILIUM

Lily of the Valley *see* CONVALLARIA

Lime *see* TILIA

Limnanthes (Poached-egg Flower) For flowering same year, seeds may be sown in March and April or later. For spring flowering, sow in autumn. All sowings out-of-doors, 0·15 cm ($\frac{1}{16}$ in) deep.

Limonium (syn. *Statice*) Annuals by seeds sown in February and March in pans of light soil, lightly covered, and germinated in a temperature of 16°C (60°F), or out-of-doors in March and April where they are to flower. Perennials by seeds sown as for annuals, or root cuttings in autumn or winter.

Linaria (Toadflax) Seeds of annuals are sown out-of-doors in a sunny spot 0·15 cm ($\frac{1}{16}$ in) deep, in April for summer flowering, and in August for spring flowering. Seeds of perennials may be sown in March and April in boxes or pans, or out-of-doors. Also increased by division in April.

Lindelfolia Seeds sown out-of-doors in a sandy soil, or division of the roots in spring.

Ling *see* CALLUNA

Linum (Flax) Seeds sown out-of-doors in April, or in pans, pots or boxes in a cold frame; shade until seeds germinate. Soft-wood cuttings of *L. perenne* and varieties root easily in sandy soil in a cold frame or under a propagator or glass jar. These are taken in June and July. *L. arboreum* and *L. flavum* can be treated in the same way.

Lion's Tail *see* LEONOTIS

Lippia (Lemon-scented Verbena) Increases readily by soft-wood cuttings in July, inserted in sandy soil in a propagating frame or under a propagator or glass jar, or under mist.

Liquidambar Seeds, which are best when imported, should be sown in a prepared, sheltered bed out-of-doors, covered 1·2 cm ($\frac{1}{2}$ in) deep, in early spring. Half-ripe cuttings 7·5–10 cm (3–4 in) long, taken with a heel, inserted in a warm propagating case in July, or under mist. Layering can be undertaken in October and November. Choose layers made from the current year's growth; give them a slight twist 15 cm (6 in) from the tip. Work peat and sand in the soil where the layers are to be pegged down; lift rooted layers the following autumn.

Liridodendron (Tulip Tree) Seeds sown out-of-doors in March. Lift and plant out seedlings eighteen months afterwards. Grafting also in March in heat under glass. Scions 12·5–18 cm (5–7 in) long are grafted on to one-year-old seedling stocks and placed in an open propagating case where there is bottom heat.

Liriope Division is the best method of propagation; divide the rhizomatous roots in March and April. Seeds can be sown in a sandy compost in March, placed in a cool greenhouse or frame.

Lithospermum Half-ripe cuttings in July under a propagator or glass jar, or in a cold frame, in a sandy peaty soil.

Lobelia Seeds sown in March in a warm greenhouse with a temperature of 13°C (55°F). Cuttings from *L. cardinalis* and its varieties may be taken in June and inserted under a propagator or glass jar; or division in March.

Lobster Claw *see* CLIANTHUS

Loganberry *see* RUBUS

London Pride *see* SAXIFRAGA

Lonicera (Honeysuckle) Cuttings 10 cm (4 in) long of current year's wood, inserted in a propagating case in June, July or August; or hard-wood cuttings 23 cm (9 in) long in October. *Lonicera nitida*, the well-known hedge plant, roots easily from hard-wood cuttings in March or December. Seeds may be sown in boxes in March in a cold frame. *L. japonica* and its varieties can be layered in summer or autumn.

Loosestrife *see* LYSIMACHIA

Loosestrife, Purple *see* LYTHRUM

Loquat *see* ERIOBOTRYA

Love-in-a-mist *see* NIGELLA

Love-lies-bleeding *see* AMARANTHUS

Luculia Half-ripe cuttings taken in early spring or early summer and inserted in a warm propagating frame where there is bottom heat, or under mist.

Lunaria (Honesty) Seeds sown 1·2 cm (½ in) deep in

April out-of-doors; thin out to 15 cm (6 in) apart.

Lungwort   *see* PULMONARIA

Lupin   *see* LUPINUS

Lupinus (Lupin)   Seeds sown out-of-doors in April and May or in boxes or pans in a light sandy soil in February and March under glass. Basal heel cuttings in early April inserted in a cold frame filled with pure sand or a very sandy mixture. Shade at first but once cuttings are erect give plenty of light except bright direct sunshine. Or division in early March. *L. arboreus* (Tree Lupin), half-ripe cuttings of sideshoots taken with a heel, in July or August and inserted under a propagator, glass jar, or in a cold frame. Seed can be sown in January and February in pans or boxes filled with John Innes seed compost and placed on a greenhouse bench with a little bottom heat, or sown out-of-doors in April and May. They can also be sown in a frame in February, small quantities in pans, pots or boxes.

Lychnis (syn. *Viscaria*)   Annuals and perennials may be sown out-of-doors in March and April where they are to flower, or in pans or boxes under glass in February and March. Division in early autumn or spring.

Lycium (Box-thorn)   Seeds sown in pots or boxes in John Innes seed compost, under glass in March. Almost-ripened cuttings 15 cm (6 in) long in September, inserted in a prepared shady border out-of-doors; or layers in October.

Lysichitum (Skunk Cabbage)   Seeds should be sown as soon as ripe, under damp conditions; sow in John Innes seed compost and keep the pot or pan in a tray of water. Division in October and November, or January and February – not later as the plants start coming into leaf very early.

Lysimachia (Loosestrife and Creeping Jenny)   Division in April or October. Cuttings of young shoots in March and April inserted in sandy soil under glass. Seeds sown in pots or boxes in March in a cold frame.

Lythrum (Purple Loosestrife)   Seeds sown in April in a prepared bed out-of-doors, or division of roots in March. Half-ripe cuttings in July inserted in an unheated frame.

Macleaya (syn. *Bocconia*) (Plume Poppy) Cuttings of sideshoots, taken in early summer, root readily if inserted in a frame or beneath a large glass jar and shaded. Suckers can be lifted at mid-summer, potted and kept in a frame until they are planted out in the following spring, or can be divided in September and replanted.

Madwort *see* ALYSSUM

Magnolia Seeds are best sown as soon as gathered. Clean the seeds of their mucilaginous coats, and sow in pans or pots filled with a mixture of equal parts loam, leafmould, fibrous peat and sharp sand with the addition of broken crocks to keep the compost open. Place a few seeds in each container and cover with light sandy soil to a depth of 1·2 cm ($\frac{1}{2}$ in). Germination takes from twelve to eighteen months. Place pots or pans in a cold frame – never use heat before germination. When seed germinates and seedlings appear above the ground, place in a heated greenhouse. Pot off singly and plant out when quite young in April. Heel cuttings 7·5–12·5 cm (3–5 in) long, in late June and July, prepared from current year's half-ripe sideshoots and inserted in pure sand in a temperature of 21–24°C (70–75°F). Layering is also a useful method for increasing magnolias. It can be done in early spring. The portion of branch that is inserted in the ground should have a thin layer of the bark slithered away for 15–30 cm (6–12 in). The prepared branch is then pegged down with two layering pegs. Layers will take two years to form a root system.

Mahonia Seeds of *M. aquifolium* (syn. *Berberis aquifolium*) should be gathered as soon as ripe, stratified and sown in the spring. Or division of sucker growths in early spring. *M. japonica*, half-ripe cuttings of current year's growth 15 cm (6 in) long in early July. Remove some of the bottom leaves only. Insert in sandy soil in a close frame. Leaf cuttings will root readily (I have read) if taken with a heel; remove the bottom three pairs of leaves and insert cuttings in sandy soil in a close frame or under mist in summer.

Mallow *see* MALVA

Malope   Seeds sown 1·2 cm ($\frac{1}{2}$ in) deep in March and April, out-of-doors where they are to flower. Thin seedlings to 23 cm (9 in) apart.

Malus (Crab Apple)   Selected garden varieties of crab apple are budded or grafted on stocks of the common crab apple *Malus pumila*. Budding is carried out in July, see Chapter Five for methods. Seed of species can be sown in spring after they have been stratified.

Malva (Mallow)   Seeds sown in March and April, 1·2 cm ($\frac{1}{2}$ in) deep in sandy soil in pans, placed in a cold frame.

Manettia   Half-ripe cuttings taken in May and June, inserted in sandy compost in a moist propagating frame, with bottom heat.

Manuka   *see* LEPTOSPERMUM

Maple   *see* ACER

Maranta (Prayer Plant)   Division in spring; pot in a compost of 1 part loam, 2 parts peat and 1 part coarse sand, and place in a temperature of 10°C (50°F).

Marigold, African and French   *see* TAGETES

Marigold, Marsh   *see* CALTHA

Marigold, Pot   *see* CALENDULA

Marjoram   *see* ORIGANUM

Marrow   *see* CUCURBITA

Marsh Marigold   *see* CALTHA

Marvel of Peru   *see* MIRABILIS

Mask Flower   *see* ALONSOA

Matricaria inodora (Scentless Mayweed)   Sow seeds under glass in March in a temperature of 13–16°C (55–60°F). Plant out seedlings in May or sow seeds out-of-doors in April where the plants are to flower.

Matthiola (Stocks)   Ten-week and Intermediate stocks by seeds, sown 0·15 cm ($\frac{1}{16}$ in) deep in pans in light soil, in a temperature of 16°C (60°F), in March. Plant out in May. *M. bicornis* (Night-scented), seeds sown out-of-doors in April. Brompton stocks, seeds sown out-of-doors or in a cold frame, 0·3 cm ($\frac{1}{8}$ in) deep in a prepared seed bed in July. Transplant in

September.

Maurandia (syn. *Maurandya*) Sow seeds under glass in February and March, in a temperature of 18–21 °C (65–70°F); grow on in a lower temperature until April, planting out in May. For winter flowering under glass, sow in April and May and grow on.

Maurandya *see* MAURANDIA

May Apple *see* PODOPHYLLUM

Mayweed, Scentless *see* MATRICARIA

Meconopsis Seeds of the Blue Poppy, *M. betonicifolia*, and other species are best sown as soon as gathered or in autumn or spring. The former is better, but it is a good plan to sow two batches, one in autumn and another in spring. Sow on the surface of the soil in pots, pans or boxes filled with a mixture of $\frac{2}{3}$ sharp sand, and $\frac{1}{3}$ granulated peat moss or powdered leaf-mould, and place in a warm greenhouse with a moist atmosphere. As soon as seedlings germinate, remove to a cooler spot and a more airy position and prick off into boxes. Pay careful attention to watering and shading. Some species like *M. cambrica*, the Welsh Poppy, will seed themselves where the seeds drop.

Medlar *see* MESPILUS

Megasea *see* BERGENIA

Melissa (Balm) *M. officinalis*, cuttings taken during the summer and inserted in sandy soil in a greenhouse, or under a large glass jar; or division of roots in spring or autumn.

Mentha (Mint) The green and variegated species of mint can be divided in autumn or spring and planted where they are to be permanently established. *M. requienii*, the minute-leaved alpine mint divide in March or April.

Mentzelia (syn. *Bartonia*) Seeds are best sown in April, in a sunny border where they are to flower, as seedlings do not transplant satisfactorily.

Menyanthes (Buckbean) Cuttings inserted in mud during the summer.

Mertensia Division of roots in April or in early Septem-

195

ber. Seeds sown as soon as ripe in pots or boxes in a frame.

Mesembryanthemum *see* LAMPRANTHUS

Mespilus (Medlar) Cleft grafting in April, or shield budding in July, on to stocks of hawthorn, pear or quince; scions should be chosen from the previous summer's growth.

Mexican Orange Blossom *see* CHOISYA

Michaelmas Daisy *see* ASTER

Michauxia Sow seeds in pans, covering them to 0·15 cm ($\frac{1}{16}$ in), in April and May in a cold greenhouse.

Mignonette *see* RESEDA

Milfoil *see* ACHILLEA

Milla uniflora *see* BRODIAEA

Mimosa *see* ACACIA

Mimulus (Musk) Seeds sown in early spring in a light soil in pans, under glass. Prick off when ready. Or cuttings of young shoots inserted under glass in spring.

Mint *see* MENTHA

Mirabilis jalpa (Four o'clock Plant, Marvel of Peru) Sow seeds under glass in February and March, 0·6 cm ($\frac{1}{4}$ in) deep in a temperature of 18–21°C (65–70°F).

Miscanthus Division of clumps in April.

Mistletoe *see* VISCUM

Mock Orange *see* PHILADELPHUS

Moluccella (Bells of Ireland) Sow seeds in early March, covering them to a depth of 0·3 cm ($\frac{1}{8}$ in), under glass in a temperature of 16–18°C (60–65°F). Or sow seeds out-of-doors in late March and April.

Monarda (Bergamot) Seeds sown in March in pots or boxes and placed in a cold frame or greenhouse – named varieties are best divided in autumn or spring.

Monkshood *see* ACONITUM

Monstera (Swiss Cheese Plant) Stem cuttings can be taken at any time of the year and inserted in a sandy compost in a temperature 21–27°C (70–80°F).

Montbretia *see* CROCOSMIA

Moraea Sow seeds as soon as ripe in a temperature of

13–16°C (55–60°F). Remove offsets any time from September to January, which is their normal planting time.

Morina (Whorl Flower)   Seeds can be sown as soon as ripe or in March, in a sandy compost to which peat and leaf-mould have been added. Or division of offsets in August and September. As plants resent disturbance, seed is to be preferred.

Morisia   Root cuttings about 2·5 cm (1 in) long taken in June and inserted in a 7·5-cm (3-in) deep pan, filled with 3 cm (1¼ in) of sand at top, plus soil and crocks, and plunged in an ash bed in a cold frame.

Morning Glory   *see* PHARBITIS

Morus (Mulberry)   Hard-wood cuttings of the current year's growth 25–30 cm (10–12 in) long taken with a heel, inserted out-of-doors in December.

Mother-in-Law's Tongue   *see* SANSEVIERIA

Mother-of-Thousands   *see* SAXIFRAGA

Mountain Ash   *see* SORBUS

Mountain Avens   *see* DRYAS

Muhlenbeckia   Hard-wood cuttings of current year's growth 5–10 cm (2–4 in) long, taken in October and November, inserted in a cold frame. Or division of established plants in March and April.

Mulberry   *see* MORUS

Mullein   *see* VERBASCUM

Muscari (Grape Hyacinth)   Increased readily by dividing clusters of bulbs when transplanting during September and October. Also increased readily by seeds sown in a greenhouse or frame in spring, but it may be a few years before seedlings flower.

Musk   *see* MIMULUS

Mutisia   Half-ripe cuttings in May and June taken with a slight heel and inserted in a propagating frame with bottom heat, or under mist. Seeds can be sown in greenhouse or frame in spring, but it may be a few years before seedlings flower.

Myosotis (Forget-me-not)   Seeds sown thinly 0·3 cm

($\frac{1}{8}$ in) deep, out-of-doors in May and June. Transplant seedlings to their final place by end of September and October.

Myrica (Bog Myrtle, Sweet Gale)   Seeds sown as soon as ripe in autumn, under glass. Half-ripe cuttings 5–7·5 cm (2–3 in) long in July and August, inserted in a propagating frame or under mist; or hard-wood cuttings inserted in sandy soil out-of-doors, in September and October.

Myrobalan Plum   *see* PRUNUS CERASIFERA

Myrtle   *see* MYRTUS

Myrtus (Myrtle)   Half-ripe cuttings taken with a heel in July and August. Insert in pure sand in propagating case, or under propagator or glass jar, or under mist, with gentle bottom heat.

Nandina domestica (Flowering Bamboo of Japan)   Half-ripe cuttings in late July and August, taken with a heel or severed at a node, 10–12·5 cm (4–5 in) long, and inserted in a sandy-peaty compost in a propagating case with bottom heat, or under mist. Hard-wood cuttings in November and December, inserted in a sandy-peaty compost in cold frame.

Narcissus (Daffodil)   Division of bulb clusters. One old bulb will often produce two others. Remove these when bulbs are lifted in July. Seeds may also be sown as soon as perfectly ripe in a gritty soil, but seedlings will take a number of years to flower and may differ widely from their parents.

Nasturtium   *see* TROPAEOLUM

Navelwort   *see* OMPHALODES

Nectarine   *see* PRUNUS

Neillia   Half-ripe cuttings in July and August 5–7·5 cm (2–3 in) long, inserted in normal cutting compost in a propagating frame with bottom heat or under mist. Or hard-wood cuttings in November 10–15 cm (4–6 in) long, inserted in normal cutting compost in a cold frame.

Nemesia   Seeds sown 0·15 cm ($\frac{1}{16}$ in) deep in pans of light soil in temperature of 13–16°C (55–60°F), in March or September, in a cool greenhouse for spring flowering. Alternatively sow seeds out-of-doors in mid-May where plants are to flower.

Nemophila Seeds sown, in September or March, out-of-doors where they are to flower. Cover seed to a depth of 0·15 cm($\frac{1}{16}$ in).

Nepeta (Catmint) Half-ripe cuttings in June and July in a cold frame, or division in March.

Nerine Division of bulb clusters when repotting in June to August. Seeds sown as soon as ripe, one seed in each 7·5-cm (3-in) pot filled with John Innes seed compost, placed in a warm propagating frame, where they will easily germinate, but it may be some years before seedlings flower.

Nerium (Oleander) Half-ripe cuttings of sideshoots 7·5–10 cm (3–4 in) long, taken with a heel or severed beneath a node, in August and September, inserted in a peaty sandy compost in a propagating frame with slight bottom heat; or ripened shoots inserted in bottles of water in late September, and placed in a propagating frame with bottom heat.

Nertera (Coral-berried Duckweed) Seeds sown as soon as ripe, or in February and March, using a peaty mixture in gentle heat under glass. Or division of the roots in March and April; pot on the divisions in a peaty, sandy compost and place in a warm greenhouse.

Nicandra (Apple of Peru) Sow seeds under glass in March in John Innes seed compost, covering 0·3 cm ($\frac{1}{8}$ in) deep, in a temperature of 16–18°C (60–65°F). Or sow seeds in April out-of-doors where they are to flower.

Nicotiana (Tobacco) Seeds may be sown in January to March in pans on the surface of light sandy soil, lightly covered with sandy soil and germinated in a temperature of 10–13°C (50–55°F). Prick off seedlings as soon as they can be handled.

Nierembergia Seeds of the annual species should be sown in May out-of-doors where they are to flower. *N. caerulea* (syn. *N. hippomanica*), half-ripe cuttings inserted in sandy soil under large glass jars in June and July. *N. repens* (syn. *N. rivularis*), division in April.

Nigella (Love-in-a-mist)   Seeds sown 0·3 cm ($\frac{1}{8}$ in) deep in March, out-of-doors where the plants are to flower.

Norfolk Island Pine   *see* ARAUCARIA EXCELSA

Nymphaea (Water-lily)   Division of the crowns in May. Old plants should be lifted and the mud washed off them; divide the tubers with a strong sharp knife. At least 7·5–10 cm (3–4 in) of tuber should be left attached to each crown. Other roots are cut back and surplus leaves removed before being replanted. The divisions must be weighted down with a turf (if not in a basket) – otherwise they will float to the surface.

Oak   *see* QUERCUS

Obedient Plant   *see* PHYSOSTEGIA

Oenothera (Evening Primrose)   Seeds of hardy annual species may be sown 0·15 cm ($\frac{1}{16}$ in) deep in patches out-of-doors in April; half-hardy species under glass in March, seedlings being planted out in May.

Biennial species are raised from seeds sown in a shady spot out-of-doors from April to July, seedlings being transplanted in September and October.

Perennial species by seeds sown in March and April, in light soil in boxes in a frame or cool house. Also by summer cuttings under a propagator or glass jar, or in a frame, or by division in autumn or spring.

Old Man's Beard, Traveller's Joy   *see* CLEMATIS

Olea (Olive)   Half-ripe cuttings, sideshoots, 5–10 cm (2–4 in) long, taken with a heel in July and August, inserted singly in small pots in a compost of 1 part loam, 1 part peat and 2 parts sand. Place in propagating frame with bottom heat or under mist. Alternatively sow seeds in February and March in a compost of 3 parts peat to 1 part sand and place in a warm greenhouse.

Oleander   *see* NERIUM

Olearia   Half-ripe cuttings of *O. haastii* 10–15 cm (4–6 in) long, inserted in sandy soil in a cold frame in late August or early September.

Other species by half-ripe cuttings in July and August

under a propagator or under mist. Hardwood cuttings inserted in a cold frame or beneath a propagator or glass jar.

Olive *see* OLEA

Omphalodes (Navelwort) Sow seeds under glass as soon as ripe in March. *O. cappadocica* division of roots in March or September when they can either be potted or planted out in a prepared site where they are to flower. *O. luciliae*, division in June; pot and keep in a shaded close frame until rooted. Once established in a garden, self-sown seeds will germinate freely.

Onion, Ornamental *see* ALLIUM

Orach *see* ATRIPLEX HORTENSIS

Orange *see* CITRUS

Origanum (Marjoram) Seeds can be sown as soon as ripe or in March, under glass. Cuttings taken in summer from basal shoots, inserted in a sand and peat mixture in a cold frame or beneath a large glass jar. Division of roots in March or October.

Ornithogalum (Star of Bethlehem) Division of offsets in September and October, planted 8·7 cm (3½ in) apart.

Osmanthus Half-ripe cuttings 7·5–10 cm (3–4 in) long in July and August, inserted in a compost of peat and sand in a propagating frame or under mist. Or layer in autumn.

Osmarea Half-ripe cuttings in May and June inserted in a propagator with medium bottom heat, or under mist.

Ostrowskia Sow seeds as soon as ripe in John Innes seed compost, in pots, placed in a cold frame. Or take root cuttings 7·5 cm (3 in) long in September and insert in a sandy cutting compost in pots or boxes, placed in a cold frame.

Othonopsis cheirifolia Seeds can be sown in spring in John Innes seed compost in pots or pans placed in an unheated greenhouse or frame. Soft to half-ripe cuttings are better, however, taken in September and inserted in John Innes cutting compost in pots, pans or boxes and placed in an unheated frame, or a few beneath a large glass jar.

Oxalis Seeds of annuals sown 0·15 cm ($\frac{1}{16}$ in) deep in a sandy soil out-of-doors in March, or under glass. Seeds of perennials sown in March, out-of-doors or under glass; or division of bulb clusters in March and April. The separated offsets may be potted or planted out.

Pachysandra Cuttings of half-ripe shoots in July and August inserted in a sandy-peaty compost in boxes, or direct into the soil in a cold frame. Division in March.

Paeonia (Peony) Seeds of the hardy herbaceous perennials can be sown as soon as ripe (September). Sow seeds 0·3 cm ($\frac{1}{8}$ in) deep in pans, pots or boxes filled with sandy soil; place in a cold frame. Seedlings may take some years to flower and are likely to differ considerably from their parents, except where they are species. Division of roots in March and April is a better method for garden varieties. Be certain that each portion divided has an eye, i.e. a new shoot. Cuttings made from young eyes taken from the base of the plants in spring may also be used. Cut each eye out with a wedge-shaped piece of flesh and place in a cold frame filled with a mixture of half peat and half sand. Keep closed for a few weeks until rooted. *P. suffruticosa* (Tree Peony), is increased by grafting on to rootstocks of *P. albiflora* and *P. officinalis*. Use a wedge-shaped graft (see p. 125) and do the work in April, starting the newly grafted plants in a propagating frame in a cool house.

Pampas Grass *see* CORTADERIA

Pansy *see* VIOLA

Papaver (Poppy) Seeds of annual poppies may be sown out-of-doors in March, April or September where they are to flower. When 2·5 cm (1 in) high, thin out to 30 cm (12 in) apart.

For perennial poppies such as *P. orientale*, sow seeds out-of-doors in March and April, or take root cuttings in boxes filled with sandy soil and placed in a cold frame in the autumn or winter.

Paradisea liliastrum (St Bruno's Lily) Seeds sown in spring in John Innes seed compost in pans or pots, placed in an unheated frame or greenhouse. Division in spring, if the roots are not too old, because with older roots rotting can occur.

Parochetus Sow seeds in February and March in sandy soil; cover seed very lightly and place in a cold frame. Soft-wood cuttings in July and August inserted in pots or pans filled with a sandy compost and placed in a cold frame. Or division of plants in March.

Parrotia Seeds sown in a sandy compost in pots or pans, in autumn or spring, and placed in a cold frame. Layering in April.

Parrot's Bill *see* CLIANTHUS

Parthenocissus (Virginia Creeper) Soft-wood or half-ripe cuttings, 7·5–10 cm (3–4 in) long, taken during June to August and inserted in a sandy cutting compost, either singly in small pots or several in a larger one, and placed in a shaded, warm propagating frame, or under mist. Alternatively hard-wood cuttings, in September and October, inserted in a peat-sand mixture in a cold frame.

Pasque Flower *see* PULSATILLA

Passiflora (Passion Flower) Heel or nodal cuttings 10–15 cm (4–6 in) long, inserted in a sandy soil in a propagating frame in July or under a propagator or glass jar, or under mist. Seeds should be removed from the Passion Flower fruits as soon as they are thoroughly ripened and stored in a dry place until February or March, when they can be sown 0·6 cm (1¼ in) deep in pots or pans filled with John Innes seed compost, and germinated in a temperature of 10–13°C (50–55°F).

Passion Flower *see* PASSIFLORA

Paulownia Seeds sown in John Innes seed compost in pans or boxes, in February and March, in a cold frame. Or take root cuttings 3·7–5 cm (1½–2 in) long and insert in a sandy compost in boxes about 7·5 cm (3 in) deep, in autumn or early spring.

Peach *see* PRUNUS

Pear *see* PYRUS

Pearl Bush *see* EXOCHORDA

Pearl Grass *see* BRIZA

Pearly Everlasting *see* ANAPHALIS

Pelargonium (Ivy-leaved, Regal, Scented-leaved and Zonal 'geraniums') Heel or nodal cuttings, made from short jointed shoots approximately 7·5 cm (3 in) long, and taken in August and September. Insert six in a 12·5-cm (5-in) pot or four in a 7·5-cm (3-in) pot in a compost of loam, peat and sand. Place in a cool house or frame. Keep in a temperature of 10°C (50°F) throughout the winter.

Penstemon Seeds may be sown 0·15 cm ($\frac{1}{16}$ in) deep in pots or pans in light soil, in a temperature of 13–18°C (55–65°F), in March. Prick out and pot, or transplant seedlings out-of-doors in May. Varieties are best increased by nodal cuttings, inserted in sandy soil in a cold frame or under a propagator, in August. Protect from frost in winter. Shrubby kinds may also be increased by half-ripe cuttings in August.

Peony *see* PAEONIA

Peperomia Seeds sown in spring in a moist, warm, temperature. Cuttings of short pieces of stem, each with a leaf attached, taken in spring and inserted in a sandy, peaty compost in a warm propagating frame; they must, however, have ample air or there is a tendency for the cuttings to damp off.

Periploca (Silk Vine) Division of the rootstock in spring; or half-ripe cuttings in July, inserted in sandy soil in a cold frame, under a large glass jar, or under mist, with bottom heat; or by layering ripened shoots in October.

Periwinkle *see* VINCA

Pernettya Species may be increased by seeds sown in a peaty soil in February and germinated in a cold house or frame. Named or selected forms, by cuttings 5–7·5 cm (2–3 in) long, inserted in a sand and peat mixture in a closed frame, or under a propagator or glass jar, in August and September; also by layering or division in spring.

Perovskia (Russian Sage) Cuttings of half-ripe shoots taken in July and inserted in a sandy compost in a propagating frame.

Peruvian Lily *see* ALSTROEMERIA

Petrea racemosa (Purple Wreath) Cuttings of firm

young shoots, taken in spring or summer, and inserted in sandy soil in a propagating frame with a temperature of 18°C (65°F).

Petunia Seeds sown, in March, in well drained pots or pans filled with light, sandy soil and germinated in a temperature of 18°C (65°F); prick off seedlings as soon as they can be handled and plant out at the beginning of June. Double varieties are best increased by cuttings in autumn or spring, though they may be raised from seed.

Phacelia Seeds sown out-of-doors in spring and on through the summer, 0·3 cm ($\frac{1}{8}$ in) deep, in a sunny border where they are to flower.

Pharbitis (syn. *Convolvulus purpurea, C. major* and *Ipomoea purpurea*) (Morning Glory) Annuals, seeds sown 0·3 cm ($\frac{1}{8}$ in) deep in a temperature of 18°C (65°F), in March. Quite a successful method is to sow three seeds in each 5-cm (2-in) pot and later put them into 20- or 23-cm (8- or 9-in) pots, once the roots have filled the small pots. Use a rich loamy compost. Soft-wood cuttings of perennials in March, or half-ripe cuttings in August, inserted in a sandy peat compost and placed in a propagating frame in a temperature of 24–30°C (75–85°F), or under mist.

Philadelphus (Mock Orange) Half-ripe nodal cuttings 10 cm (4 in) long, made from the tips of sideshoots and inserted in a propagating case filled with sand in June and July; or under mist. Bottom heat is necessary. Alternatively hard-wood cuttings 30 cm (12 in) long, inserted out-of-doors in a prepared bed in autumn.

Phillyrea Cuttings of half-ripe shoots, 7·5–10 cm (3–4 in) long, taken in July and August and inserted in a sandy soil in a cold frame, or under mist; or layering in autumn.

Philodendron Cuttings in May and June, made from shoots each having three joints, inserted in a sandy, peaty compost in a propagator, where there is a warm, moist atmosphere, with a temperature of 24–27°C (75–80°F). Or layering in similar temperature as for cuttings.

205

Phlomis (Jerusalem Sage) Cuttings of half-ripe shoots taken in June and July and inserted in a sand and peat compost, in a propagating frame. Or hard-wood cuttings, cut below a node and inserted in a cold frame in October and November.

Phlox Annual seeds sown in March, 0·3 cm ($\frac{1}{8}$ in) deep in pans filled with light soil, in a frame or greenhouse with gentle bottom heat; or out-of-doors in a prepared bed in autumn. *P. paniculata* (Perennial phlox) can be increased from basal cuttings in March, April or summer, inserted in a cold frame, or deep boxes filled with sand or a sandy cutting compost. But, as eelworm is common in the perennial phlox, I favour root cuttings in autumn, see pp. 82 and 91. Choose the best roots and cut into lengths of 5–7·5 cm (2–3 in). Lay these pieces horizontally in boxes filled with a compost of 2 parts loamy soil, 2 parts leafmould or peat, and 1 part sand; cover with fine soil and place in a cold frame or greenhouse. Plant out in spring.

Variegated varieties of phlox like 'Harlequin' and 'Norah Leigh' need to be propagated from cuttings, for if increased from root cuttings, the foliage becomes green. *P. subulata* (Alpine phlox) and its varieties by heel or nodal cuttings about 2·5 cm (1 in) long, inserted in sand or sandy soil in cold frame, or under a propagator or glass jar, June and July.

Phormium Seeds sown in March in John Innes seed compost in a greenhouse. Or division of roots in April.

Photinia Cuttings of half-ripe shoots, taken with a heel and inserted in a compost of 2 parts peat, 1 part loam and 1 part sand, and placed in a propagating frame, in July and August. Or hard-wood cuttings taken in October and November and inserted in a cold frame.

Phygelius (Cape Figwort or Cape Fuchsia) Half-ripe cuttings, with or without heel, 5–7·5 cm (2–3 in) long, inserted in propagating case, or under hand-light or glass jar, in July and August. Seeds may be sown in March in gentle heat.

Physalis (Cape Gooseberry or Chinese Lantern) Seeds sown in a cold frame or greenhouse in March or out-of-doors in April. Division of roots in March and April.

Physostegia (Obedient Plant) Division in spring. Or cuttings of firm, young shoots 5–7·5 cm (2–3 in) long, inserted in a sandy compost in pots and placed in a cold frame.

Phyteuma Seeds sown as soon as ripe in August and September, in a well drained gritty compost. Division of roots in spring.

Phytolacca (Pokeberry) Seeds sown out-of-doors in spring or autumn. Division in October or March.

Pieris Cuttings of half-ripe shoots, taken with a heel and inserted in a sand and peat compost in pots, placed in a cold frame, under a large glass jar, or under mist, in August and September. Seeds sown in a peat and sand compost or a soilless seed compost in March, in a cold frame. Or layering in autumn.

Pilea (Artillery Plant) Take soft stem cuttings 7·5–10 cm (3–4 in) long in May, insert in a mixture of equal parts sand and peat in 7·5-cm (3-in) pots and place in a propagator, in a temperature of 18–21 °C (65–70°F).

Pineapple *see* ANANAS

Pineapple Flower *see* EUCOMIS

Pinks *see* DIANTHUS

Piptanthus laburnifolius (syn. *P. nepalensis*) Sow seeds in March in John Innes seed compost in a cool house. It also seeds itself freely around the parent plant. Half-ripe cuttings can be taken in July and August, and inserted in a heated propagating frame in a greenhouse.

Pittosporum Half-ripe cuttings 5–7·5 cm (2–3 in) long in June and July, inserted in a propagating frame with bottom heat. Seeds may be sown, in March and April, in a soilless compost in pans or pots, placed in a cool greenhouse or frame.

Plagianthus Half-ripe cuttings of sideshoots 7·5–10 cm (3–4 in) long, in July and August, taken with a heel and inserted in a sandy-peaty mixture in a propagating frame, or under mist. Alternatively, layers of the previous year's growth, given a slight twist about 20–25 cm (8–10 in) from the end of the shoot, and put down in early spring, root readily.

Plagiorhegma   *see* JEFFERSONIA

Plane Tree   *see* PLATANUS

Plantain Lily   *see* HOSTA

Platanus (Plane Tree)   Seeds sown in a sandy compost, in boxes in a cold frame in spring. Layering in spring; after giving the shoot a twist 20 cm (8 in) from its end, insert it in the ground. Or hard-wood cuttings of the current season's growth, taken in December, 20–25 cm (8–10 in) long, inserted in a sheltered bed out-of-doors.

Platycodon (Balloon Flower)   Seeds sown in February and March in John Innes seed compost in a cold frame. Or sow seeds as soon as ripe. Division of roots in March or October.

Plectranthus   Cuttings root readily inserted in peaty soil, any time from April to June, under glass in gentle heat.

Plum   *see* PRUNUS

Plumbago capensis   Heel cuttings 7·5 cm (3 in) long in late March and April, inserted singly in small pots filled with sandy soil and placed in a close propagating frame.

Plume Poppy   *see* MACLEAYA

Poached-egg Flower   *see* LIMNANTHES

Pocket-handkerchief Tree   *see* DAVIDIA

Podophyllum (May Apple)   Division of roots in March and April. Or seeds sown as soon as ripe or in March, in a cold frame.

Poinsettia   *see* EUPHORBIA PULCHERRIMA

Pokeberry   *see* PHYTOLACCA

Polemonium (Jacob's Ladder)   Seeds sown under glass in February and March or out-of-doors in April. Division of roots in March or September.

Polianthes (Tuberose)   Although it can be propagated from offsets, they seldom ripen sufficiently and it is wiser to obtain imported stock of this bulb-like tuberous-rooted perennial.

Polyanthus   *see* PRIMULA

Polygonatum (Solomon's Seal)   Division in March or

October. Seeds are best sown as soon as ripe in John Innes seed compost, either in pots or direct in a cold frame; grow on seedlings for two to three years before planting them permanently.

Polygonum Herbaceous perennials by seed sown in a cool house or frame in spring, or by division of roots in March or April. *P. vaccinifolium*, half-ripe cuttings inserted under a propagator or glass jar, or in a cold frame, in July and August. *P. baldschuanicum* (Russian Vine), heel cuttings of half-ripe wood in July and August, inserted in a warm sand frame. Or hard-wood cuttings 30 cm (12 in) long, inserted in an open sheltered border or frame out-of-doors, from November to February.

Pomegranate *see* PUNICA

Poncirus (syn. *Aegle sepiaria*) Seeds sown in March, in pans of John Innes seed compost; cover the seeds 0·6 cm ($\frac{1}{4}$ in) deep, place pots or pans in a cold greenhouse or frame. Half-ripe cuttings in June and July, inserted in sandy soil in a propagating frame.

Pontederia cordata (Water Plantain) Easily increased by division in April; plant the divisions with at least 15−30 cm (6−12 in) of water above the crowns.

Poor Man's Orchid *see* SCHIZANTHUS

Poplar *see* POPULUS

Poppy *see* PAPAVER

Populus (Poplar) Hard-wood cuttings 30 cm (12 in) long root very easily, if inserted in a prepared bed or border out-of-doors in November and December.

Portugal Laurel *see* PRUNUS

Portulaca Seeds sown in March in pans filled with light soil. As the seed is very minute, only a bare sprinkling of sand is required as covering. Place in temperature of 18°C (65°F). Alternatively, sow seeds out-of-doors in April where plants are to flower.

Potentilla Seeds sown in February in cold frame or greenhouse. With shrubby kinds choose half-ripe cuttings of the current year's growth in late summer, and insert in cold frame, or under a propagator or glass jar, or under mist. Alpine and perennial kinds by division in March and April.

Pot Marigold   *see* CALENDULA

Prayer Plant   *see* MARANTA

Prickly Thrift   *see* ACANTHOLIMON

Primula   Seeds of *P. malacoides* may be sown in April, May and June (to give a succession of flowers) in a cold frame or house, in pans, pots or boxes filled with light soil. Sow seeds of *P. obconica* in March and May in a temperature of 13°C (55°F). Sow seeds of *P. sinensis* in April in a temperature of 16–18°C (60–65°F). Cover very lightly. Seeds of *P. elatior* (Polyanthus) may be sown in March under glass in light soil. Seeds should be covered lightly with leafmould or chopped moss. Or seeds may be sown out-of-doors in July and August in a shaded place. Prick off when large enough to handle. *P. auricula* is increased by offsets, side growths or top of plant, taken off when repotting, which is immediately after flowering; or by seeds as for *P. elatior*. Alpine species and varieties can be divided after flowering. *P. denticulata* (Drumstick Primula) seeds can be sown in July and August out-of-doors in a shady spot or increased by root cuttings in December and January, in the same way as phlox root cuttings.

Privet   *see* LIGUSTRUM

Prophet Flower   *see* ARNEBIA

Prunella   Division in March or September.

Prunus   This genus includes the flowering and fruiting almonds, apricots, cherries, nectarines, peaches and plums. To perpetuate selected varieties true to name they must be propagated by budding or grafting on to the appropriate stocks. Budding is the more usual method. This operation is performed in July and August and grafting in March – see Chapter Five. Seeds after being stratified may be sown in spring in pots of sandy soil. Species can be increased satisfactorily in this way. Laurel, Cherry (*P. laurocerasus*), and Portugal (*P. lusitanica*), hard-wood cuttings in October, inserted in a north border. Myrobalan Plum (*P. cerasifera*), hard-wood cuttings in winter inserted out-of-doors.

Pulmonaria (Lungwort)   Division in autumn or spring.

Pulsatilla vulgaris (Pasque Flower)   Seeds are best sown as soon as ripe, June and July, in pans or pots filled with John Innes seed compost. Cover the pans with a piece of glass and place in a cold frame. Root cuttings 2·5 cm (1 in) long, taken in July and August and inserted vertically in pans filled with equal parts loamy soil, leafmould and coarse sand, in a cold frame.

Pumpkin   *see* CUCURBITA

Punica granatum (Pomegranate)   Half-ripe cuttings inserted in sandy-peaty compost in closed propagating frame with medium bottom heat in June and July, or under mist.

Purple Heart   *see* SETCREASEA

Purple Loosestrife   *see* LYTHRUM

Purple Rock Cress   *see* AUBRIETA

Purple Wreath   *see* PETREA

Puschkinia (Striped Squill)   Seeds sown as soon as ripe in boxes or pans filled with John Innes seed compost, barely covering the seeds. Keep moist and place in frame. During the winter keep pans or boxes out-of-doors. Alternatively, division of bulbs; lift after flowering and replant, or they can be stored in a cool place and planted in early autumn.

Pyracantha (Firethorn)   Cuttings of current year's growth, about 10 cm (4 in) long, with a heel, inserted under a propagator or in a closed frame in late July and August, or under mist. Hard-wood cuttings inserted in a cold frame in autumn.

Pyrethrum   *see* CHRYSANTHEMUM COCCINEUM

Pyrus (Pear)   Selected garden varieties of pears are increased by budding in July and August, or grafting in March on to the appropriate stocks, see p. 103.

Quaking Grass   *see* BRIZA

Quercus (Oak)   Sow acorns as soon as ripe in open ground in autumn. Named varieties are grafted under glass in March.

Quickthorn   *see* CRATAEGUS

Quince, Common *see* CYDONIA OBLONGA

Quince, Japanese *see* CHAENOMELES

Ramonda Seeds may be sown as soon as ripe or in March in gentle heat. Division or leaf cuttings in April.

Ranunculus Tuberous rooted species are propagated by separating the offsets from older tubers in autumn, when they are lifted for storing. Seeds of all species may be sown as soon as ripe in pans or boxes of sandy soil. They germinate best in a cold frame or cool greenhouse. Herbaceous species are increased by division in early autumn or spring.

Raoulia Sow seeds under glass in March or divide roots in July and August.

Raspberry *see* RUBUS

Red-hot Poker *see* KNIPHOFIA

Red Valerian *see* KENTRANTHUS

Reed Mace *see* TYPHA

Regal Geranium *see* PELARGONIUM

Rehmannia Seeds sown in May in a cool greenhouse or frame, in pots or pans of ordinary seed compost, only just covering the seed. Cuttings can be taken in autumn, inserted in pans or pots and placed in a propagating frame with bottom heat. Or root cuttings in autumn.

Reinwardtia To obtain cuttings, plants should be cut back after flowering, about the end of March. Take cuttings in April and May 5–7.5 cm (2–3 in) long and insert in sandy soil in a propagating frame with bottom heat.

Reseda (Mignonette) Seeds sown in pans, 0.6 cm ($\frac{1}{4}$ in) deep, in March in cool house, or out-of-doors in April and May. For early flowering in the greenhouse, sow in August and keep in a cool house throughout the winter.

Rhamnus (Buckthorn) Seeds sown out-of-doors in autumn. Hard-wood cuttings in September and October. Layering in early spring.

Rhaphiolepis Half-ripe cuttings of sideshoots about 5 cm (2 in) long taken with a heel, inserted in pots filled with a sandy, peaty compost and placed in a propagating frame with gentle bottom heat, in September and October, or under mist.

Rheum (Rhubarb) Both ornamental species grown in the flower border and rhubarb in the vegetable plot can be propagated from seed sown out-of-doors in April. Division of both in February and March.

Rhipsalidopsis *see* SCHLUMBERGERA

Rhododendron These beautiful shrubs may be propagated by seed, grafting, cuttings and layers. Seeds are best sown during the middle of January, in pans half-filled with crocks and peat moss litter and placed in a propagating frame in a temperature of 13–16°C (55–60°F). Seeds must be perfectly cleaned. Seed should only have the thinnest of sand covering.

Grafting of selected garden varieties is done in February, the common *R. ponticum* being used as a rootstock. Saddle grafting is the method usually employed and grafted plants are started in a propagating frame with bottom heat, see p. 123.

Cuttings of the hardy hybrids can be taken from mid-July to August and inserted in a sandy, peaty compost under mist. The alpine rhododendrons root fairly readily under a propagator or under mist from July to October, or even November.

Layering is best undertaken in July and August, see pp. 56–8. This method is very satisfactory for amateurs.

Azalea seeds sown when ripe or in March in a sandy, peaty mixture, in pans or pots placed in a cool house. Half-ripe cuttings in July inserted under a propagator or under mist. Grafting in February and March in heat on to rootstocks of *R. luteum* (syn. *Azalea pontica*). Layering in May and June. Half-ripe cuttings of *R. indicum* and *R. simsii* (*Azalea indica*) in July and August, inserted in equal parts sand and peat, in pots or pans placed in a propagating frame or hand-light in a greenhouse or under mist.

Rhodohypoxis Offsets of the corm-like rhizomes can be removed and divided in March and replanted.

Rhoicissus Cuttings of sideshoots 7·5 cm (3 in) long taken in April and May, inserted in a peat and sand compost in pots or pans, and placed in a propagating frame in a temperature of

16–18°C (61–64°F).

Rhubarb   *see* RHEUM

Rhus cotinus   *see* COTINUS

Rhus cotinoides   *see* COTINUS

Rhus glabra (Smooth Sumach) and *R. typhina* (Stag's-horn Sumach)   Root cuttings 3·7–5 cm (1½–2 in) long, inserted in sandy-peaty compost in boxes in early spring; or by suckers detached in autumn.

Rhyncospermum   *see* TRACHELOSPERMUM

Ribes   This genus embraces the flowering currant *R. sanguineum*, other ornamental species, and the fruiting kinds. All may be readily increased by sowing seed in pans or boxes filled with light sandy soil, in March, and placing in a cold frame. To keep selected garden varieties of red, white and black-currants and gooseberries and flowering currants true to type, hard-wood cuttings must be taken in late autumn or winter. They should be about 30 cm (12 in) long, of the current year's wood, and inserted in the open ground. Blackcurrants do not have their buds removed from the portion inserted in the ground; the remainder do.

Richardia   *see* ZANTEDESCHIA

Ricinus (Castor Oil Plant)   Seeds sown 1·2 cm (½ in) deep, singly in 7·5-cm (3-in) pots, filled with light sandy soil and plunged in a close propagating case with bottom heat in March and April. When young plants are large enough to handle, pot off into 12·5- or 15-cm (5- or 6-in) pots, using a compost of sandy loam, leafmould and a little rotted manure.

Robinia (False Acacia)   Seeds sown in spring in pots or pans in a cool greenhouse; or suckers taken up in the autumn. Species such as *R. hispida* and *R. kelseyi* are grafted on to *R. pseudoacacia*, also varieties such as 'Frisia', in spring out-of-doors.

Rochea   To obtain suitable cuttings, cut back plants in April; when basal shoots appear, take cuttings 7·5–10 cm (3–4 in) long, insert them in a coarse sand and peat mixture in pots or pans in 13–16°C (55–61°F). Keep them fairly dry.

Rock Jasmine *see* ANDROSACE
Rock Cress *see* ARABIS
Rock Rose *see* CISTUS
Rodgersia Seeds sown in John Innes seed compost in pots or pans and placed in a cold frame in March. Or division of roots in March.
Romneya (Californian Tree Poppy) Root cuttings 5 cm (2 in) long, inserted singly in small pots in a sandy compost, in a propagating frame with bottom heat, in the spring.
Rosa Seeds of rose species germinate readily if sown in boxes of sandy soil in January and February after being stratified.

Selected garden varieties of hybrid tea, floribunda roses, etc., are usually increased by budding in July and August on to rootstocks of various rose species, such as *Rosa canina, R. rugosa* and *R. laxa*, but some kinds, particularly ramblers, and also certain hybrid teas, shrub roses, and species, can be increased by cuttings of ripe shoots of the current year's growth inserted outdoors in September or early October, or by layering around midsummer.

Miniature roses, cuttings in July and August, preferably with a heel, or firmer cuttings in October. Insert in sandy soil in a cold frame, or under a propagator or glass jar.
Rosemary *see* ROSMARINUS
Rosmarinus (Rosemary) Half-ripe cuttings inserted in a cold frame in sandy soil in August, or under mist.
Rose of Sharon *see* HYPERICUM CALYCINUM; for Canada and USA *see* HIBISCUS
Rubber Plant *see* FICUS ELASTICA
Rubus (Blackberry, Loganberry and Raspberry) Seeds of species sown in pots or pans in spring germinate readily in a cool house or cold frame. Stem or root cuttings are also successful if inserted in a cold frame in July and August. *R. giraldianus* and *R. odoratus* may be divided in early spring. Raspberries by suckers removed in autumn. The ornamental rubus with long trailing stems (such as the whitewash brambles) layer

naturally where their tips touch the ground, and this method may also be used for fruiting blackberries and loganberries, see p. 52.

Rudbeckia Annuals (hardy and half-hardy) can be sown under glass in February and March, in a temperature of 16–18°C (60–65°F) and planted out in May. Or seeds can be sown in pots or boxes in a cold frame in March and April. Perennial species can be treated similarly – though with the latter, division of plants is preferable, in autumn or spring.

Rue *see* RUTA

Ruellia Cuttings of young shoots 5–7·5 cm (2–3 in) long, inserted in sand or in equal parts sand and peat, in a warm propagating case in April.

Ruscus (Butcher's Broom) Division of rootstock in spring. Seeds should be stratified during the winter and sown in pots or seed trays the following spring, in a cold frame.

Russian Sage *see* PEROVSKIA

Russian Vine *see* POLYGONUM BALDSCHUANICUM

Ruta (Rue) Seeds sown in March and April in pots or pans, in a cold frame. Half-ripe cuttings in July and August, 7·5–10 cm (3–4 in) long, inserted in sandy, peaty compost in a cold frame, or a few under a large glass jar, or under mist.

Sage, Common *see* SALVIA OFFICINALIS

Sagittaria (Arrowhead) Division of roots in March and April.

St Bernard's Lily *see* ANTHERICUM

St Bruno's Lily *see* PARADISEA

St John's Wort *see* HYPERICUM

Saintpaulia (African Violet) Seeds are minute and require similar treatment to begonias. Germinate in a temperature of 16–18°C (60–65°F). They can also be increased by pegging down leaves on the surface of sandy soil in a close propagating frame, preferably with bottom heat.

Salix (Willow) Hard-wood cuttings root very easily. They may be anything from 30 to 90 cm (1 to 3 ft) long and are inserted out-of-doors in autumn or early winter. Alpine kinds like *S. retusa* may be increased by heel cuttings 2·5–5 cm (1–2 in) long, inserted in a warm propagator in June and July, or under mist.

Salpiglossis The time to sow the seeds will depend on when the plants are wanted to bloom. Autumn-sown seeds in a cold frame for plants to flower in January to March, January sowing for April and May flowering in a temperature of 18–21 °C (65–70°F).

Salsify *see* TRAGOPOGON

Salvia *S. fulgens* and its varieties by cuttings, either in July and August, or in March, rooted in a close propagating frame, preferably with bottom heat; or by seeds sown in January and February in a temperature of 18°C (65°F). *S. horminum*, seeds sown out-of-doors in April 0·3 cm (⅛ in) deep. *S. officinalis* (Common Sage), cuttings strike readily in late summer, or old bushes may be layered by having their centres filled with soil in autumn or March, when the branches will form roots by the following May or June. *S. sclarea* (Clary) and its variety *turkestanica*, although both are short-lived perennials, are best grown as biennials from seeds sown in a cold frame in May and June. *S. × superba* (syn. *S. virgata nemorosa*), by division in spring or autumn.

Sambucus (Elder) Hard-wood cuttings 15–20 cm (6–8 in) long, with a heel, any time from December to February, inserted in sandy soil in open ground out-of-doors. Or seed, which should be cleaned by being rubbed through sand and stored until ready for sowing in spring, out-of-doors.

Sanguinaria (Blood Root) Seeds sown out-of-doors in April, or division of roots in March or October. Half-ripe cuttings inserted in sand soil in a frame, or under a glass jar.

Sanguisorba (Burnet) Seeds sown in pans or pots in John Innes seed compost, in March and April, in a cold frame. Or division of roots in March and April.

Sansevieria trifasciata laurentii (Mother-in-Law's Tongue) Rooted sucker growths or offsets can be removed by being cut off with a sharp knife and potted singly in 7·5-cm

(3-in) pots in John Innes potting compost No 1, in spring. Or take leaf cuttings about 7·5 cm (3 in) long, made from sections of the leaf, and insert them in a sandy-peaty mixture in a warm propagating frame. But these cuttings will produce only mottled green foliage – without the yellow margins.

Santolina (Lavender Cotton or Cotton Lavender)   Half-ripe cuttings, with or without a heel, inserted in a cold frame, or under a propagator or glass jar, in July and August.

Saponaria (Soapwort)   Annual species by seeds sown in April where they are to flower. Perennial species such as *S. ocymoides* by division in spring, or half-ripe cuttings inserted under a propagator or glass jar, or in a cold frame, in July and August.

Sarcococca   Division in spring, or hard-wood cuttings taken in October and November, 7·5–10 cm (3–4 in) long, inserted in a cold frame.

Sarracenia   Division in March and April, inserted in pots in a mixture of 3 parts peat, ½ part each of sphagnum moss and charcoal, and placed in a warm propagating frame. Or seed can be sown on top of a similar compost in pots or pans in a temperature of 13–16°C (55–60°F).

Satin Flower   *see* SISYRINCHIUM

Satureia (Savory)   *S. hortensis* (Summer Savory) is an annual and seed should be sown out-of-doors in April. *S. montana* (Winter Savory), division in early spring or half-ripe cuttings 5–7·5 cm (2–3 in) long, inserted in a sandy-peaty mixture in a propagating frame.

Savory   *see* SATUREIA

Saxifraga   The majority of this genus are readily increased from seeds sown in spring in pans filled with gritty soil and placed in a cold frame. Encrusted section may also be propagated by offsets potted up singly, or used as cuttings and inserted in a sandy gritty soil; the mossy section by cuttings inserted in a cold frame in the summer or by division in autumn or spring. London Pride, *S. × urbium* (syn. *S. umbrosa*, the name it has been known by for many years), division in spring or autumn.

S. stolonifera (Mother-of-Thousands) can be propagated by the tiny plantlets that appear on the stems or runners, and potted on into small pots.

Scabiosa Seeds of *S. caucasica* can be sown in March and April. Sow thinly in pans or boxes filled with a sandy compost in an unheated frame or greenhouse. Prick off seedlings when large enough to handle. Named varieties of *S. caucasica* can be propagated by basal cuttings in spring, inserted in a sandy compost in a frame or greenhouse. Annual scabious by seeds sown out-of-doors in light soil, in April for flowering end of summer or in August to flower the following summer. Alternatively, sow in gentle heat in March in shallow boxes or pans in light soil and prick off for planting out in May.

Scabious *see* SCABIOSA

Scarborough Lily *see* VALLOTA SPECIOSA

Scented-leaved Geranium *see* PELARGONIUM

Schisandra Half-ripe cuttings 5–7·5 cm (2–3 in) long taken with a heel in August, inserted in a sand and peat compost and placed in a propagating frame with medium bottom heat. Or under mist.

Schizanthus (Butterfly Flower or Poor Man's Orchid) Seeds can be sown for greenhouse cultivation from July to September in pans or boxes in a cold frame. For early flowering sow in July. For flowering out-of-doors sow in February and March in a temperature of 18°C (65°F) or in the open in April.

Schizophragma Half-ripe cuttings 3·7–5 cm (1½–2 in) long taken with slight heel, or nodal cuttings in July and August, in a sand and peat compost placed in a propagating frame in temperature of 16°C (60°F); or under mist. Layering in spring. Seeds sown under glass in February and March.

Schizostylis (Kaffir Lily) Division of clumps in spring, but leave four to six shoots to each divided portion.

Schlumbergera (syn. *Zygocactus truncatus*) The Christmas Cactus is now known as *S. buckleyi* and forms of this and *S. truncatus* are available; also the Easter Cactus, previously known as *S. gaertneri* but now named *Rhipsalidopsis gaertneri*.

All are easily increased from cuttings in spring, summer or winter. Cut off pieces of stem, each having three or four stem joints, leave them two or three days in a shady place, afterwards inserting them in pots filled with a well drained compost by pushing each cutting in to the length of the lowest joint. Keep the cuttings away from sunlight until they are rooted. Winter cuttings will need to be rooted in a warm propagator.

Scilla (Squill) Seeds sown 0·15 cm ($\frac{1}{16}$ in) deep in boxes of sandy soil in a cold frame in September. It will be three to four years before the seedlings flower. Separation of offsets from parent bulbs in August. This is the better method.

Scirpus lacustris (Common Bulrush) Division of roots in March, replanted in shallow water at the edge of a pool.

Scolopendrium *see* FERNS

Scotch Flame Flower *see* TROPAEOLUM SPECIOSUM

Scrophularia (Figwort) Cuttings can be taken in autumn and inserted in sandy soil in a cold frame. Or division in spring.

Scutellaria Seeds sown in March and April in pots or pans and placed in a cold frame. Or division in March.

Sea Buckthorn *see* HIPPOPHAË

Seaholly *see* ERYNGIUM

Seakale *see* CRAMBE

Sedum (Stonecrop) Sow seeds of annual species in April where plants are to flower.

Alpine species may be divided in late summer or autumn. *S. spectabilis* may be divided in spring, or take stem cuttings in July and August and insert in a sandy compost under a hand-light or glass jar.

Selaginella Division of plants annually when repotting in spring.

Sempervivum (Houseleek) Division in spring or early summer, or seeds sown in well drained pans or boxes in spring and germinated in a cold frame or greenhouse.

Senecio cineraria (syn. *Cineraria maritima*) Half-ripe cuttings with or without a heel, inserted under a glass jar or hand-light in July and August. *S. cruentus* (Cineraria), in order to maintain a succession of flowers, several sowings must be made. The first batch of seeds is sown in April for December flowering, the second early in June for January and February, and the third in July for late spring flowering. Where only one sowing is made the best time is May. Seeds sown in April will require a little warmth. Other sowings can be in a cool house. Seeds are only just covered with very fine soil. Good drainage of pots and pans is essential. Young seedlings or plants should not be exposed to full sun in the summer months. *S. greyi*, often listed as *S. laxifolius* in catalogues, half-ripe cuttings in August or early September, inserted in a cold frame or under a propagator or glass jar, or under mist.

Setcreasea (Purple Heart) Cuttings about 10 cm (4 in) long taken in June to August; insert about five cuttings around the inside edge of an 8·7-cm (3½-in) pot, filled with John Innes compost No. 2, and place under propagator in a greenhouse, or beneath a large glass jar.

Shasta Daisy *see* CHRYSANTHEMUM MAXIMUM
Shrimp Plant *see* BELOPERONE
Sidalcea Division of roots in autumn and spring. Seeds germinate freely if sown in April in light sandy soil in pans or boxes in a cold frame, but seedlings may differ considerably from their parents.

Silene Annual species by seeds sown out-of-doors in August and September, and transplanted when large enough to handle. Alternatively, sow in April and transplant in May to flowering quarters. Perennial species by division in spring.

Silk Oak *see* GREVILLEA
Silk Vine *see* PERIPLOCA
Sinningia (Gloxinia) Seeds sown in well drained pans from January to March or in the autumn, in a mixture 1 part each of peat moss, loam and silver sand, all sifted fairly fine. Fill within 1·2 cm (½ in) of the rim. Mix seed with fine silver sand to ensure even distribution. Do not cover the seed with soil, but a very light covering of silver sand; place a pane of glass and a sheet of paper over each pan. Keep in a moist

atmosphere in a temperature of 18–24°C (65–75°F) and shade until germination takes place, after which gradually remove both glass and paper. Prick out about 2·5 cm (1 in) apart as soon as first two leaves are developed, to prevent damping-off. Gloxinias can also be propagated by cuttings of shoots 2·5– 5 cm (1–2 in) long, by young leaves with stalks, or by leaf cuttings slit through the mid-ribs. Compost for all is a sandy-peaty mixture and all should be inserted under a propagator in a temperature of 18–21°C (65–70°F).

Siphonosmanthus delavayi *see* OSMANTHUS

Sisyrinchium (Blue-eyed Grass, Satin Flower) Seeds sown in pots or pans in March in a cold frame. Self-sown seedlings also appear freely once plants are established. Division of roots in March or September.

Skimmia Seeds can be sown in February in a cold frame. Half-ripe cuttings 7·5–10 cm (3–4 in) long inserted in sandy compost in a warm propagating frame or under mist in July and August. Or hard-wood cuttings inserted in a cold frame in October and November.

Skunk Cabbage *see* LYSICHITUM

Smilacina racemosa (False Solomon's Seal) Division of plants in October.

Smilax Seeds sown in spring in pots or pans in a temperature of 16–18°C (60–65°F), or division of plants in March.

Smithiantha Division of the rhizomes broken into 5-cm (2-in) pieces and potted in a compost of 2 parts peat, 2 parts leafmould and 1 part sand; set the rhizomes about 2·5 cm (1 in) deep, in the spring. Leaf cuttings in May and June inserted in a sandy-peaty mixture in a propagating frame in a temperature of 16–24°C (70–75°F).

Smoke Tree *see* COTINUS

Smooth Sumach *see* RHUS GLABRA

Snapdragon *see* ANTIRRHINUM

Snowberry *see* SYMPHORICARPOS

Snowdrop *see* GALANTHUS

Snowdrop Tree *see* HALESIA

Snowflake *see* LEUCOJUM

Snow in Summer *see* CERASTIUM

Snowy Mespilus *see* AMELANCHIER

Soapwort *see* SAPONARIA

Solanum Seeds of *S. capsicastrum* (Christmas Cherry) are sown in pans of light sandy soil in a temperature of 18°C (65°F) at the end of February or early March. *S. crispum* and *S. jasminoides*, half-ripe cuttings in summer inserted in a propagating case, under a propagator or glass jar, or under mist. *S. wendlandii* is increased by cuttings of young shoots, which root readily in sand in early spring if inserted in a close propagating case with bottom heat, or under mist.

Soldanella Division of the roots in June, pot them in a compost of 3 parts soil, 1 part leafmould, and 1 part sharp sand, with a little mortar rubble added to the mixture, and place in a cold frame.

Solidago (Golden Rod) Division of roots in October or March.

Solidaster Division of roots in October or March.

Solomon's Seal *see* POLYGONATUM

Sophora Seeds sown in February and March. Soak the seeds in hot water for about two hours, then sow in pots or pans in ordinary seed compost, and place in a propagating frame with bottom heat. Half-ripe cuttings taken with a heel in August and placed in a propagating frame with bottom heat, or under mist.

Sorbaria Seeds sown in March in pans or boxes of light soil and lightly covered, and placed in a cool greenhouse or cold frame, germinate freely. Half-ripe cuttings in July and August in a cold frame or under mist. Hard-wood cuttings in autumn inserted in open ground or frame.

Sorbus (a) Forms of *S. aria* (Whitebeam) can be budded in July and August or grafted in March on to rootstocks of *S. aria*. (b) Selected garden forms of *S. aucuparia* (Mountain Ash) can be budded in July and August or grafted in March on to rootstocks of *S. aucuparia*. Seeds of species of sorbus can be sown in spring after the seed has been stratified.

Southernwood *see* ARTEMISIA

Spanish Broom *see* SPARTIUM

Spanish Iris *see* IRIS XIPHIUM

Sparaxis Seeds sown in August and September in pots or pans under glass in a temperature of 16°C (60°F), or division of offsets in early spring.

Sparmannia (African Hemp, House Lime) Plants cut back in April after flowering will produce plenty of soft cuttings which root readily inserted in equal parts sand and peat in a temperature of 16°C (60°F).

Spartium (Spanish Broom) Seeds sown in boxes, under glass, in February and March.

Spider Flower *see* CLEOME

Spider Wort *see* TRADESCANTIA

Spindle Tree *see* EUONYMUS

Spiraea Seeds sown in March in pans or boxes of light soil and covered lightly, and placed in a cool greenhouse or cold frame, germinate freely. Half-ripe cuttings in July and August in a cold frame, or hard-wood cuttings in autumn inserted in open ground or frame. For the pinnate leaved types such as *S. aitchisonii*, take root cuttings in autumn. Many can also be increased by division of suckers in autumn or spring, e.g. *S. menziesii*.

Spurge *see* EUPHORBIA

Squill *see* SCILLA

Stachys grandiflora (syn. *Betonica grandiflora*) *S. lanata* (Lamb's Ear) and *S. l.* 'Silver Carpet', can all be increased by division in spring or autumn.

Stachyurus Layering in June and July; layer shoots of the current year's growth – no incision in the shoot is necessary, simply a slight twist before inserting layers in the ground. Half-ripe cuttings, about 7·5 cm (3 in) long taken with a heel and inserted in a warm propagator in July and August, or under mist.

Stag's-horn Sumach *see* RHUS TYPHINA

Staphylea (Bladder Nut) Seeds sown thinly in pots or pans in light sandy seed compost, under glass in February and March. Half-ripe cuttings 5–7·5 cm (2–2 in) long inserted in a sandy peat compost beneath a large glass jar or under mist in July and August. Layering in July and August; make a slight twist about 15 cm (6 in) from tip of shoot of current year's growth, and insert layers in the ground.

Star of Bethlehem *see* ORNITHOGALUM

Statice *see* LIMONIUM

Stephanandra *S. incisa*, division in early spring. *S. tanakae*, layers in summer; give young shoots a slight twist, insert in a very sandy soil and peg down layers. Hard-wood cuttings 23–30 cm (9–12 in) long inserted in a prepared bed in October.

Stephanotis Cuttings taken in spring of the current year's wood, inserted in a compost equal parts sand and peat in a propagating frame with bottom heat of about 18°C (65°F), or under mist.

Sternbergia Remove offsets once the foliage is really withered, which is usually about April.

Stewartia Seeds should be sown in February and March in a warm greenhouse. Soak the seed for a few days before it is sown, thinly, in boxes filled with 2 parts loam and 1 part each of peat and sand. Give only a light covering and shade until germination. Pot off seedlings when 5 cm (2 in) high and grow on in a greenhouse until established. Cuttings of half-ripe shoots about 7·5 cm (3 in) long with a heel can be taken in early June and inserted in a compost of 3 parts sand and 1 part each of peat and loam, or a soilless cutting compost, in pots placed under a large glass jam jar or under mist.

Stipa Seeds sown in March in a warm frame or greenhouse or out-of-doors in April. Or division of roots in March and April.

Stocks *see* MATTHIOLA

Stokesia Seeds sown in March and April in pans filled with peat or a soilless seed compost and placed in an unheated greenhouse or frame. Or division of rooted offsets in March; or unrooted offsets can be used as cuttings taken in March.

Stonecrop *see* SEDUM

Stransvaesia Seeds should be gathered in autumn and stratified in peat and sand. Sow the seeds in spring out-of-doors in a prepared bed. Leave seedlings in seed bed until autumn. Layering can be undertaken in the autumn; plant out layers the following autumn. Or half-ripe cuttings with a heel, 7·5–10 cm (3–4 in) long in July, inserted in equal parts peat and sand in a propagator with gentle bottom heat.

Strawberry This useful fruit may be increased by seed, division or layering, the latter method being the most useful. Seeds may be sown in boxes in the autumn and allowed to remain out-of-doors throughout the winter. Seed boxes are placed under glass in a frame or house in spring, which ensures even germination.

Division of roots after dry summers (when runners are not very free) is a useful method. Layering is described in Chapter Three, see p. 54.

Strawberry Tree *see* ARBUTUS

Strelitzia (Bird of Paradise Flower) Division of plants or removal of rooted sideshoots or suckers in spring, when they should be potted and kept in a warm moist atmosphere. Or seeds can be sown in pans filled with a soilless or ordinary seed compost in a temperature of 18–21°C (65–70°F) in March and April.

Streptocarpus Seeds are sown either in January or July for autumn and summer flowering respectively. Sow in pots. Soil must be fine and exactly level, and thin sowing is essential. A temperature of 18°C (65°F) is required and pots should be kept in the dark until germination takes place. Propagation may also be effected by leaves inserted in pots filled with osmunda fibre and placed in a close frame with bottom heat in March, see p. 89. Or division in March.

Streptosolen Cuttings of sideshoots about 7·5 cm (3 in) long in March and April inserted in a peaty-sandy soil in a propagating frame in a temperature of 16–18°C (60–65°F), or under mist.

Striped Squill *see* PUSCHKINIA

Styrax Seeds can be sown as soon as ripe, in pans of loamy soil, and placed out-of-doors. Alternatively, seeds can be sown in February and March in pans filled with 1 part each of peat and sand and 2 parts loam, and placed in a warm greenhouse. Layering in autumn using the current season's growth. Or half-ripe cuttings in June and July in a propagating frame or under mist.

Sumach *see* RHUS

Summer Cypress *see* KOCHIA

Summer Hyacinth *see* GALTONIA

Summer Starwort *see* ERINUS

Sunflower *see* HELIANTHUS ANNUUS

Sun Rose *see* HELIANTHEMUM

Sweet Bay *see* LAURUS

Sweet Gale *see* MYRICA

Sweet Pea *see* LATHYRUS

Sweet Rocket *see* HESPERIS

Sweet William *see* DIANTHUS

Swiss Cheese Plant *see* MONSTERA

Sycamore *see* ACER

Sycopsis Cuttings of ripened side-shoots taken in October or early November and inserted in a cold frame.

Symphoricarpos (Snowberry) Hard-wood cuttings inserted out-of-doors in autumn, or by removal of suckers.

Symplocos Cuttings of half-ripe shoots taken with a heel in July. Use a mixture of loam, peat and plenty of sand and insert in a propagating frame with gentle bottom heat, shading until rooted, or under mist. Seeds can be sown in February and March, in pans filled with a compost 1 part sand and 2 parts peat; place in a cold house or, better still, a cold frame, then twelve months later (in January) place the pans in a warm greenhouse.

Synthyris Division after flowering.

Syringa (Lilac)   Species may be increased by seeds, which germinate readily if sown in pans or boxes in March and placed in a cold north frame. Selected garden varieties do not breed true from seed but may be increased by heel cuttings 7·5 cm (3 in) long of the current year's wood, inserted in a warm sand frame in June or under mist, or by firmer cuttings in August inserted in a cold frame. Layering is also a successful method if carried out in the spring; or garden varieties may be budded on to *Syringa vulgaris* in early summer. With budded plants there is always the problem of suckers arising from the stock plant. To avoid this remove all buds or eyes from the base of the stock. This will help to reduce sucker growth. Even with layered plants these, too, will produce an abundance of suckers, and if they are not removed, indifferent flowering can result – when removing the suckers scrape away the soil and pull them off rather than cutting them. *S. velutina* (syn. *S. palibiniana*), soft-wood cuttings in spring inserted in a warm propagating case or under mist.

Tagetes (Marigold, African and French)   Seeds may either be sown 0·15 cm ($\frac{1}{16}$ in) deep in pans or boxes filled with light soil and germinated in a temperature of 16°C (60°F) in March, or be sown out-of-doors at the end of April where plants are to flower.

Tamarisk   *see* TAMARIX

Tamarix (Tamarisk)   Hard-wood cuttings 20–30 cm (8–12 in) long inserted out-of-doors in a bed of rather light, sandy soil in October.

Tanacetum (Tansy)   Seeds can be sown out-of-doors in spring, or division in spring or autumn.

Tansy   *see* TANACETUM

Tea Tree   *see* LEPTOSPERMUM

Tecoma   *see* CAMPSIS

Teucrium (Germander)   The alpine forms can be propagated from cuttings taken in July and inserted in pots in a cold frame, or by division of roots in March. *T. fruticans*, half-ripe cuttings 5–7·5 cm (2–3 in) long in June to early August, inserted in sandy peaty soil in a propagating frame or under mist.

Thalictrum   Seed of *T. dipterocarpum* may be sown in pans or boxes of sandy soil under glass in spring. *T. dipterocarpum* 'Hewitt's Double' may be propagated by taking off new shoots, which grow some 5 cm (2 in) away from the old flowering shoot. These should be removed just before they show above ground in the spring, and then potted singly in small pots or placed in boxes filled with sandy cutting compost. Start in a cold frame.

Thermopsis   Seeds of this leguminous plant can be sown under glass in February, or out-of-doors in April. Transplant seedling plants at an early age as they resent disturbance. Division of roots in March and April.

Thrift   *see* ARMERIA

Throat Wort   *see* TRACHELIUM

Thunbergia alata (Black-eyed Susan)   Sow seeds of this half-hardy annual in February and March under glass in a temperature of 18–21°C (65–70°F). When seedlings are large enough to handle, prick off singly into small pots.

Thyme   *see* THYMUS

Thymus (Thyme)   Seeds may be sown in spring or summer in pans, pots or boxes filled with sandy soil and placed in a cold frame or out-of-doors. *T. serpyllum* and its many varieties are best increased by division after flowering. Shrubby forms such as *T. citriodorus* 'Silver Queen' and *T. nitidus* may be increased by heel or nodal cuttings in July, inserted under a propagator or in a cold frame. Thyme as used in the herb garden is easily increased by division in autumn or spring.

Tiarella   Sow seeds out-of-doors in April, or division in March.

Tibouchina   Soft-wood cuttings can be rooted in a warm propagating frame in early spring, or half-ripe cuttings can be taken during the summer and rooted without or with very little heat, or under mist. In either case use a very sandy compost.

Tidy Tips   *see* LAYIA

Tiger Flower *see* TIGRIDIA

Tigridia (Tiger Flower) Remove offsets when the bulb-like corms are lifted in autumn. Store them in sand in a dry, cool, frostproof place.

Tilia (Lime) Sow seeds in pots as soon as gathered in autumn. Plunge the pots out-of-doors throughout the winter, and remove them to a warm greenhouse in the spring. Layering in spring is a satisfactory method. Varieties can also be grafted out-of-doors in spring, using the whip and tongue method. Or chip budding, see p. 110, is successful with *T. platyphyllos rubra*.

Toadflax *see* LINARIA

Tobacco *see* NICOTiANA

Tolmiea Increased by pegging down the strawberry-like runners, so that the base of the leaves touch the soil.

Torenia Sow seeds in pots or pans in John Innes seed compost or a soilless compost, under glass in a temperature of 18–21°C (65–70°F). The seed should be sown on the surface of the compost; water by immersion.

Trachelium (Throat Wort) This half-hardy, sub-shrubby perennial is best treated as a biennial, and to keep a good supply, sow annually. Seeds can be sown in June and July in pans or pots, in very fine seed compost or a soilless one, and placed in a frame out-of-doors. Or half-ripe cuttings can be taken in July and inserted in a sandy peaty compost in a propagating frame.

Trachelospermum jasminoides (syn. *Rhyncospermum*) Half-ripe cuttings in June and July inserted in a propagating frame with a temperature of 18–21°C (65–70°F) or under mist. Layering can also be used in spring or summer.

Trachycarpus fortunei (Chusan Palm) Seeds are sown in spring in a temperature of 24–27°C (75–80°F); or division of suckers.

Tradescantia virginiana (Spider Wort) and its varieties. Seeds can be sown in pans, pots or boxes and placed in a cold house or frame in spring. Division of plants in spring or autumn. *T. blossfeldiana* and *T. fluminensis* (Wandering Jew) are readily increased from cuttings inserted in a sandy or soilless compost from April to August, under glass or in a propagating frame.

Tragopogon (Salsify)   Seeds can be sown in April and May, out-of-doors in 2·5 cm (1 in) deep drills 25–38 cm (10–15 in) apart. When seedlings are large enough to handle, thin to 20–25 cm (8–10 in) apart.

Traveller's Joy   *see* CLEMATIS

Tree of Heaven   *see* AILANTHUS

Tree Lupin   *see* LUPINUS ARBOREUS

Tree Peony   *see* PAEONIA SUFFRUTICOSA

Tree Poppy, Californian   *see* ROMNEYA

Tree Purslane   *see* ATRIPLEX HALIMUS

Tricuspidaria (syn *Crinodendron*) Half-ripe cuttings in July and August 6·2–7·5 cm (2½–3 in) long, inserted in a sandy, peaty compost in a propagating frame or under mist with bottom heat.

Trillium (Trinity Flower)   Division of their rhizomatous roots in late summer and autumn; do not allow the rhizomes to become dry before replanting. Seeds can be sown in March in a soilless seed compost or a peaty seed mixture in a cold frame.

Trinity Flower   *see* TRILLIUM

Triteleia   *see* BRODIAEA

Tritonia   Division of offsets in autumn when the montbretia-like corms are lifted after flowering. Store in a dry frostproof place and replant in March and April.

Trollius (Globe Flower)   Seeds may be sown in pans or boxes in very sandy soil in spring, but trollius seed is slow to germinate. A method that is successful is to sow in pure sand 0·6 cm (¼ in) deep, placing the pan in a saucer of water and covering with a piece of glass until germination occurs. Seed may also be sown out-of-doors in a prepared shady border in early summer. However, the simplest and most satisfactory method for the amateur is by division of roots in spring, by which selected garden varieties can be kept true to type.

Tropaeolum and **T. majus** (Nasturtium)   Seeds sown in spring out-of-doors where plants are to flower. *T. peregrinum*

(Canary Creeper), seeds can be sown in April out-of-doors where plants are to flower, or for indoor cultivation sow seeds in February and March under glass in a temperature of 13–16°C (55–60°F). Tuberous rooted kinds, sow seeds in light sandy soil in a temperature of 13–18°C (55–65°F) in spring; or cuttings of young shoots inserted in a similar temperature in spring or summer. Double flowered forms can be increased by cuttings inserted in a closed frame with bottom heat in July. *T. speciosum* (Scotch Flame Flower) and *T. polyphyllum* may be increased by seeds sown in a cold frame or house in April, or by division of roots at planting time in the spring.

Tuberose   *see* POLIANTHES

Tulip   *see* TULIPA

Tulipa (Tulip)   Removal of offsets when bulbs are lifted in July.

New varieties are raised from seed sown in spring, or better still, as soon as the seed is ripe. Sow seed 0·6 cm ($\frac{1}{4}$ in) deep, in pans filled with 7·5 cm (3 in) depth of friable soil. Cover with glass and place pans on the shady side of a hedge or wall in a cold frame. Water from base. Give protection until seeds have germinated, then put in a cold frame. Give good attention to ventilation and watering. As leaves begin to flag, the pans of seedlings may be dried off and the young bulbs turned out in June and stored in sand until planting time in October. It will be anything from four to seven years before seedling bulbs will flower.

Tulip Tree   *see* LIRIODENDRON

Tunica   Seeds can be sown in ordinary seed compost or a soilless one under glass in February, or out-of-doors in April. Cuttings in April and May inserted in a cold frame or division of roots in March and April.

Typha latifolia (Bulrush, Reed Mace)   Division of roots can be done in March or October.

Ulex (Gorse)   Species can be increased by seeds sown out-of-doors in the spring. *U. europaeus* 'Plenus' is best propagated by cuttings inserted in a propagator where there is bottom heat or under mist in July. Grow on in pots, as gorse resents transplanting from open ground.

Ulmus (Elm) Sow seeds as soon as ripe, out-of-doors in a prepared bed, covering the seed 1·2 cm ($\frac{1}{2}$ in) deep. Leave seedlings in seed bed for twelve months after germination, then line out in a nursery bed. *U. procera* (English Elm) does not produce fertile seed and it must therefore be propagated from suckers.

Umbellularia (Californian Laurel) Half-ripe cuttings of the current season's growth about 7·5–10 cm (3–4 in) long can be inserted in a propagator in July and August, or under mist.

Umbrella Plant *see* CYPERUS

Ursinia Sow seeds in March under glass in a temperature of 16–18°C (60–65°F). Prick out in the normal way. Or seeds can be sown out-of-doors in April.

Uvularia Division of the rhizomatous roots in early autumn.

Vaccinium Seeds may be sown under glass in February in a sandy-peaty mixture or finely chopped sphagnum moss, lightly cover the seeds only. Cuttings of half-ripe shoots of the current year's growth can be taken in August and inserted in a sandy-peaty mixture beneath a plastic propagating cloche, or under mist. Other means are division of roots or suckers.

Valeriana montana and **supina** Division in July and August and grown on in a cold frame during the winter. *V. phu aurea*, division of the roots in autumn.

Vallota (Scarborough Lily) Offsets of *V. speciosa* (syn. *V. purpurea*) can be removed when repotting in June and July.

Variegated Laurel *see* AUCUBA

Veltheimia Removal of offsets in September, when plants should be repotted. This is only necessary about every third or fourth year.

Venetian Sumach *see* COTINUS

Venidio-arctosis Cuttings 7·5–10 cm (3–4 in) long are made from lateral shoots and taken during August and September. Insert them in equal parts of peat and sand in a propagating frame with a temperature of 13–16°C (55–60°F). Pot on the rooted cuttings into 7·5 cm (3-in) pots using John Innes potting compost No 1. During the winter they should be placed in a temperature of 4–7°C (40–45°F). Give a final potting in March, into 10-cm (4-in) pots, in readiness for planting out in May.

Venidium   Seeds are sown under glass in a temperature of 16–18°C (60–65°F). Sow seeds in pans or pots and only barely cover the seed. Or sow seeds in May out-of-doors where the plants are to flower.

Veratrum (False Hellebore)   Division of roots in October, or March and April.

Verbascum (Mullein)   Root cuttings in autumn or winter. Sow seeds in April in a cold frame.

Verbena   Bedding varieties by seeds sown 0·15 cm ($\frac{1}{16}$ in) deep, at the end of January or in February, in pans filled with loam, leafmould and sand, and germinated in a temperature of 16°C (60°F). Prick off when two or three leaves appear.

Verbena (Lemon-scented)   *see* LIPPIA

Veronica   Alpine and herbaceous kinds either by division in early autumn or spring, or by cuttings inserted in a cold frame or under a propagator or glass jar, in spring or early summer.

Veronica (Shrubby)   *see* HEBE

Viburnum   Seeds are sown in March, but are slow to germinate, often taking up to two years. To hasten germination, plunge the pots out-of-doors in winter and place in a propagating frame in spring. Heel cuttings of most species and varieties root easily if made from half-ripe wood in June, July and August, and inserted in a close frame, or under mist. The deciduous species and varieties can also be increased from hard-wood cuttings in autumn out-of-doors. Layers may also be pegged down in June. *V. carlesii* can be rooted from cuttings but is often budded or grafted. This is done in July, on to stocks of *V. lantana*.

Vinca (Periwinkle)   A very easy plant to propagate as

234

stems root readily at the tips. Plants may also be divided in early spring. Or cuttings of current year's growth, inserted in a cold frame in October.

Viola (Pansy, Viola and Violet) Bedding violas: seeds should be sown thinly in a pan or box containing light soil, in early March. Place in a frame or greenhouse; a little heat is an advantage but not essential. Selected varieties are best increased by cuttings of young shoots taken in August from the heart of the plant. The shoots selected should not have flowered. Insert in a prepared frame or bed. Each cutting should be about 4 joints long. Division can also be effected in April.

Pansy: seeds are sown in boxes or pans filled with light sandy soil during July and August and placed in a shady part of the garden out-of-doors. Or seeds can be sown in the open ground in a prepared bed out-of-doors. Sow the seed thinly and give only a very light covering of soil; afterwards water the bed gently but thoroughly and shade from sun until germination. Cuttings may also be taken in August and September and inserted in a shaded frame. Division of old roots may be carried out in September and October but this is not a very satisfactory method.

Violet: Seeds can be sown in early autumn in light soil in a shallow box, covered with a piece of glass and placed in a cold frame or house. They can also be increased by division in spring but a far better method is to prepare cuttings from the runners produced freely in late summer and root these in sandy soil in a frame or under a hand-light. The rooted runners will be ready for planting out the following April.

Violet *see* VIOLA

Virginia Creeper *see* PARTHENOCISSUS

Viscaria *see* LYCHNIS

Viscum album (Mistletoe) To obtain berries it is necessary to grow both male and female forms fairly close together. Mistletoe is propagated by bursting the berry in spring on the underside of a youngish branch of the host plant. The glutinous substance in which the seed is enclosed will soon harden and become attached securely. A slit can be made in the bark but it is not really necessary.

Vitis (Grape) Seeds of species may be sown in March in pans and placed in a cold frame. For fruiting vines, eyes taken in December are the best form of propagation, as described in Chapter Four, p. 81. Cuttings of well-ripened shoots also root easily if made 30 cm (12 in) long in October.

Waldsteinia Division of roots in March, when they can be planted in their permanent position.

Wallflower *see* CHEIRANTHUS

Wandering Jew *see* TRADESCANTIA, ZEBRINA

Wand Flower *see* DIERAMA

Water-lily *see* NYMPHAEA

Water Plantain *see* PONTEDERIA

Watsonia Offsets can be removed when the corms are lifted in the autumn, and stored in a dry, frostproof place until April and May when they can be replanted. Or seeds may be sown in pans filled with a light rich compost, under glass in a temperature of 16°C (60°F) in February and March.

Wax Plant *see* HOYA

Weigela This genus is closely related to Diervilla. Seeds of species sown in March in a cold frame; cuttings of half-ripe shoots in late June and July, inserted in a sandy soil under a glass jar or propagator, or under mist; or hard-wood cuttings in October in a cold frame or out-of-doors.

Welsh Poppy *see* MECONOPSIS CAMBRICA

Whitebeam *see* SORBUS

Whorl Flower *see* MORINA

Willow *see* SALIX

Windflower *see* ANEMONE

Winter Sweet *see* CHIMONANTHUS

Wisteria Selected varieties are increased in March by grafting small pieces of the branch on to the roots of *W. sinensis* and placing these in a closed propagating frame with bottom heat. Half-ripe cuttings can be taken in August, inserted in a sandy, peaty compost, in a propagating frame in a greenhouse.

Layering in July is another means of propagation. Seed can be sown in March in pans filled with sandy soil and placed under glass in a cool house, but seedlings seldom flower freely and usually plants take many years before they flower at all.

Witch Hazel *see* HAMAMELIS

Woad *see* ISATIS

Woodruff *see* ASPERULA

Yarrow *see* ACHILLEA

Yucca The tops can be induced to root by trimming off half the leaves and placing the stem in a pot filled with sandy soil, standing them in a greenhouse until rooted. *Y. flaccida* is increased by division. Most species produce rhizomatous underground stems. These will form young plants when cut off and potted. Root cuttings can be inserted in sandy soil in heat in winter or early spring.

Zantedeschia (syn. *Calla, Richardia*) (Arum Lily) Arum lilies are usually increased by removing the small rhizomes or offsets at the time of repotting, in late August and September. The offsets may be grown on in large seed boxes or frost-free frames. Larger rhizomes can be cut up with a sharp knife, making sure each portion has a dormant bud attached. After offsets have been removed, or old rhizomes cut, treat the wounds with powdered charcoal and lime. Seeds can be sown 0·3 cm ($\frac{1}{8}$ in) deep, singly in small pots filled with a compost of loam, leafmould and sand, in a temperature of 18°C (65°F).

Zanthoceras Seeds when obtainable can be sown in February and March. Sow in pots filled with a compost of 2 parts soil, 1 part each of peat and sand. Cover seed very lightly and place in an open propagator where there is gentle bottom heat. Root cuttings can be taken in December and January. The roots should be inserted horizontally in pots and placed under glass.

Zauschneria Division of the plants in March or April. Soft-wood cuttings taken from the end of May to early June or half-ripe cuttings from July to September. Insert in a sandy cutting compost in a propagator. Seed can be sown in March in pots in a cold frame.

Zebrina (Wandering Jew) Cuttings root readily at the joints, inserted in a sandy compost or a soilless compost from April to August under glass. Or they can be rooted in a glass of water.

Zelkova Sow seeds in February and March in pans filled with a sandy-peaty compost or a soilless compost and place in an open case where there is medium bottom heat in a greenhouse. Whip and tongue grafting in spring out-of-doors using 15–20 cm (6–8 in) scions of previous year's wood, and grafting on to stocks of *Ulmus glabra*.

Zenobia Sow seeds in spring in pans filled with well sifted moss-peat, lightly cover seeds and place pans in a cold frame, shade heavily, and when seeds have germinated admit a little air at a time. Or layers can be put down in autumn.

Zephyranthes candida Can be increased readily from offsets. However, if a few bulbs are planted and left, they will soon naturally increase on their own.

Zinnia Seeds are sown 0·15 cm ($\frac{1}{16}$ in) deep in pans or pots filled with light soil in March and April in a temperature of about 16°C (60°F). When seedlings are large enough to handle, pot into 7·5-cm (3-in) pots in a sandy compost; place them under glass, planting them out in May.

Zonal Geranium *see* PELARGONIUM

Zygocactus *see* SCHLUMBERGERA

MANUFACTURERS AND SUPPLIERS

Soil Sterilizers and Propagators can be obtained from the following manufacturers and suppliers:

STERILIZERS
Camplex (see Simplex), Humex, Nobles and Shilton.

PROPAGATORS
Access Frames, Autogrow, Camplex (see Simplex), Eltex, Humex, Stewarts and Ward.

Access Frames   Crick, Northampton.

Autogrow   Quay Road, Blyth, Northumberland.

Camplex
Simplex of Cambridge Ltd., Horticultural Division, Cambridge, CB2 4LJ.

Eltex   George H. Elt Ltd., 48 Eltex Works, Worcester.

Humex   5 High Road, Byfleet, Surrey, KT14 7QF.

Nobles   (Wellingborough) Ltd., Wellingborough, Northants.

Shilton
Garden Products Ltd., 390 City Road, London, EC1V 2QA.

Stewarts   Plastic Ltd., Purley Way, Croydon, CR9 4HS.

Ward
George Ward (Moxley) Ltd., Baggotts Bridge, Darlston, Staffs.

❧ Index ❧

Entries in the Alphabetical List of trees, shrubs and other plants with methods of propagation, have not been indexed, except where some special mode of propagation has been mentioned. The figures in bold indicate illustrations.

Runners, layering, 52, 54–6, **55**

Saddle grafting, 118, 123–5, **124**
Scales, 73
  insertion of, 91, **90**
Scion, definition of, 101
  for root grafting, 125–7, **126**
  selection of, 116–17
Sealed seed packets, 17–18
  Bodgers, 18
  Dunns, 18
  Hurst, 18
  Suttons, 18
Secateurs, 67
Seed, 15–39
  bed, preparation of, 24
  cleaning, 17
  composts, 30–5
  covering, after sowing, 27–9, **28**
  crocking, boxes, pans and pots, 27, 29
  drainage of containers, 29
  dressing, 30
  dust-like, 18–19
  fleshy, 19, 21
  frost, effect on, 26
  hard-coated, 19, 21
  harvesting, 17
  F$_1$ hybrids, 16
  F$_2$ hybrids, 16
  hybrid vigour, 16
  oily, 19, 21
  packeting, 17
  pelleted, 18, 25–6
  plumed, 19, 21
  sealed packets, 17
  sieve, 24
  snow, effects on, 26
  So-Fresh, 18
  sowing, 21–7, 29, **22**
    cold frame, 23, 26
    cold greenhouse, 23, 26
    depth of, 23, **22**
    large seeds, **22**
    open ground, 23–4
    warm frame, 23, 26
      greenhouse, 23, 26
    temperatures for, 26

  watering after, 29
    before, 25
  storing, 17
  stratification, 19, 21, **20**
  sweet peas, 19
  types of, 18–19
  winged, 19, 21
Seedlings, pricking out, 36–9, **37**, **38**
Serpentine layering, 53–4
Sharpening stick, 115–16
Side grafting, 118, 121, 123, **122**
Sieve, 24
Snagging, 115
Snow, effects on seed, 26
So-Fresh, 18
Soft-wood cuttings, 76–7, **77**
  temperature for, 76
Soilless composts, 32–3
Soil sterilizers, 34–6, **35**
  Nobles Steam, **35**
  Shilton Gas, **35**
  Sterilizing, 34–6
    saucepan method, 34–5
  tilth, 24
Sowing, depth for seed, 23, **22**
  in frames, 26
  in greenhouses, 26–7
  large seeds, **22**
  methods, 24–7, 29
  out-of-doors, 23–4
  times, 24
  watering after, 25, 29
    before, 25
Sphagnum moss, 62–3
Splice grafting, 118, 120
Spores, see ferns, 167, **167**
Sterilizers, soil, 34–6, **35**
Sterilizing soil, 34–6
Stock and scion compatibility, 102
  definition, 101–2
Stocks, apple, 102–3
  briar cuttings for, 104
    seedlings for, 104
  cherry, 103
  clematis, 125–7, **126**
  Doucin, 102
  dwarf fruit, 102
  for budding, 104